The Theory and Practice of
Therapeutic Touch

Jean Sayre-Adams RN MA
Director and Senior Lecturer, Didsbury Trust

Stephen G Wright MBE RGN
FRCN DipN RCNT DANS RNT MSc(Nursing)
Director, The European Nursing Development Agency (TENDA); visiting Professor of
Nursing, University of Southampton

with contributions by

Francis C Biley MSc(Manchester) BNurs(Wales) RMN RGN PGCE FETCert
Lecturer in Nursing Science, University of Wales College of Medicine; Co-ordinator,
International Region of the Society of Rogerian Scholars

Michael Richardson RN BN(Hons)
Staff Nurse, Infectious Diseases Unit, University Hospital of Wales

Illustrated by

Murti Schofield BEd(drama) ACCT
Animator, Assistant Producer and Production Designer in Film and Television

Foreword by

George Lewith MA MRCP MRCGP
The Centre for Study of Complementary Medicine, Southampton

CHURCHILL LIVINGSTONE
EDINBURGH HONG KONG LONDON MADRID MELBOURNE NEW YORK
AND TOKYO 1995

CHURCHILL LIVINGSTONE
Medical Division of Longman Group Limited

Distributed in the United States of America by Churchill Livingstone Inc., 650 Avenue of the Americas, New York, N.Y. 10011, and by associated companies, branches and representatives throughout the world.

First published 1995

ISBN 0 443 05227 1

British Library Cataloguing in Publication Data
A catalogue record for this book is available from the British Library.

Library of Congress Cataloging in Publication Data
A catalog record for the book is available from the Library of Congress.

The
publisher's
policy is to use
paper manufactured
from sustainable forests

Produced through Longman Malaysia

Contents

Foreword

It is a great privilege for me to be asked to write a foreword to this text. I know both the authors and have very considerable respect for their work and achievement in integrating complementary medicine into nursing.

Philosophically and culturally, nursing – and indeed medicine as a whole – has been driven further and further down the technological road over the last 20 years. Nurses have difficulty caring and touching because this is somehow not seen as a useful task-orientated achievement. Unfortunately complementary medicine has often been seen as 'soft' and frequently there has been little good research to back up the claims made for its efficacy.

The efforts made to understand and evaluate Therapeutic Touch sets an example to us all. This technique grew from the experience of those with a background in healing. All too often within complementary medicine the development stops at this point and the therapist will begin to look inwards, but this has not been the case with Therapeutic Touch. Sensible critical questions have been asked about the technique by those working in the field. Many answers have emerged, so we now have a good body of research which has been clearly outlined in this text to support the use and value of Therapeutic Touch as an important if not essential part of the whole process of nursing. The integration of Therapeutic Touch into mainstream nursing can therefore be research led.

It is intuitively apparent to those who work within the health-care field that caring and touching are immutably bound together with the healing process. It is therefore not surprising that over the last 20 years Therapeutic Touch and other similar techniques have emerged as a counter current to the increasing mechanisation of medicine. The case of Therapeutic Touch is, however, a wonderful example to all of us working within the field of complementary medicine. As the authors so clearly describe, it takes an initial intuitive approach, develops it into a clear technique, which can be taught and practised by nurses working in very complex, high-technology environments, and then goes yet further in demonstrating clearly through the use of clinical trials that this technique is of real value. From a nurse's point of view, this allows real integration between modern medicine and some of the oldest healing processes available to us. It is a truly complementary approach to the patient, harnessing the best of all possible approaches for their condition.

1994

GL

Preface

If you don't know what to do,
just put your hands on them and love them

PANNATIER

When words fail, there can be no better way to convey to the patient that 'I care' than by touch. However, there are many ways of touching and even more ways in which the patient might respond. In the modern world of high-tech medicine and drugs, complex professional roles, attempts at cost control and higher patient throughput, patients can often become more distant from, rather than closer to, those who seek to heal them.

It cannot be coincidence that, worldwide, millions of people have turned to complementary therapists for help. If these therapies have witnessed a growth boom in recent years, there has been an equally strong backlash. Sceptics, traditionalists, religious fundamentalists and many others have conducted a rearguard action to hold back the tide. Yet demands for access to complementary therapies escalate inexorably, perhaps because they offer an intimacy of contact that is lacking in many aspects of health care. The complementary therapies all tend to focus on a close helping relationship between therapist and patient, and they offer comfort and sometimes healing when other measures have failed or are no longer appropriate. The complementary therapies also tend to focus on the individual, seeking to help the whole person and the many aspect of wellbeing, not just the disease process. For some, they may even offer a spiritual dimension that could be lacking in their lives. Complementary therapies, compared to many other health care techniques, are usually pleasant to experience and pain free – they are often used in conjunction with meditation, relaxation, music, guided imagery and so on. Thus the

focus may frequently be on putting the patient in the best condition
for self–healing to occur, rather than relying upon complex technical
interventions.

Many of the therapies also place a heavy reliance upon touch, but
what sort of touch? How does ordinary touch differ from that which
is therapeutic? What is the difference between the touch used when
washing the patients' skin or helping them to move, and that which
seeks to help and heal? Can such therapies work *with* conventional
medicine or do they go against the grain?

One technique, Therapeutic Touch (TT), has been gaining increased
attention in recent years. It relies upon the belief that people are not
solid matter to which our everyday perceptions are accustomed, but
that they are fields of energy. Through appropriate training it is
possible for us to become aware of this energy, and for each of us to
tap into that awareness. Thus we can help to repattern and balance
the energy fields of others which have become distorted through
stress or illness. It is possible, through the practice of TT, to help and
heal another by creating the context in which relaxation and maximum
health can occur. The TT practitioner does not simply transfer healing
energy to another, but participates in a highly intricate and complex
exchange, with the intention of putting the patient in the best position
for self-healing to take place.

TT does not require patients to reject established health care techniques, rather it seeks to work with them in a complementary and
supportive way. Nurses and many other health care workers find
themselves ideally placed to incorporate TT into their everyday
practice, as the many examples in this text will illustrate. It is relatively
easy to learn, requires no special equipment or drugs and costs little
more than the practitioner's time. It seems to help many people
while also being safe – the worst that can happen is nothing at all.

In response to the growing demand for more information, we have
compiled this book which will explore TT background, research,
practice and education. We are grateful to all the contributors who
have given much time and effort to this project – enabling us to
produce what will be a guide for those who seek to develop the
practice of TT themselves. Through the work of the Didsbury Trust,
a charity dedicated to the teaching and development of TT in the

UK, many thousands of nurses and other health care workers have now learned of the benefits of TT, both for themselves and for their patients.

1994   JS-A
                                                                   SGW

Authors' note

Throughout the text, for simplicity, we have referred to nurses in the generic sense, including midwives and health visitors. Other health care workers practising Therapeutic Touch (TT) need only change the word 'nurse' to their own preferred title.

Again for simplicity, we have used the word 'patient' when referring to someone receiving TT.

Acknowledgements

We are grateful to Janet Venn, Carol Horrigan, Ann Mills, Jill Fawke and Barbro Perkins for providing material used as case studies in parts of the text. Felicity Rankin, whose story appears on page 80, drew the picture on the back cover.

1

An introduction and background to Therapeutic Touch

Jean Sayre-Adams   Steve Wright

Recent thinking on the nature of energy
Concepts of human energy
The emergence of Therapeutic Touch
Why us?

■ *The real voyage of discovery consists not in seeking new lands but in seeing with new eyes.*

MARCEL PROUST

Therapeutic Touch (TT) is a modern version of one of the oldest of therapies – the use of touch as a means of comfort and healing. At the same time the understanding of why it has lasted down through the centuries is at the cutting edge of many modern scientific disciplines, quantum physics for example.

In this chapter we will be looking at a general overview of TT which traces some of the history of the laying on of hands. Although Chapter 2 will explore the research done on TT, some historical background is also included to offer some additional dimensions.

Evidence of the use of touch as a method of healing can be traced back 15 000 years to drawings found on the walls of caves. Early Eastern philosophical and religious literature, Greek mythology, ancient Egyptian artefacts, native American Indian myths, all contain some mention of touch as a means used to heal. Evidence for the use of laying on of hands to heal human illness in ancient Egypt is found in the Ebers Papyrus dated 1552 BC. Four centuries before Christ, the Greeks used touch therapy in their Asklepian temples for healing the sick. The writings of Aristophanes detail the use of laying on of hands in Athens to restore a blind man's sight and return fertility to a barren woman. Hippocrates (460–377 BC) recorded, 'It has often appeared, while I have been soothing my patients that there was a singular property in my hands to pull and draw away from the affected parts, aches and diverse impurities by laying my hand upon the place and by extending my fingers toward it' (cited in MacManaway & Turcan 1983).

The Bible has many references to laying on of hands for both medical and spiritual applications (see Ch. 2). Healing was considered as much part of the early Christian ministry as preaching or administering the sacraments. In Europe, the healing ministry was also carried on as the 'royal touch' and lasted until the reign of William IV. Many of these earlier attempts at healing were based upon a belief either in the power of Jesus, the king or a particular healer. Even now, many healers do so in the name of a spirit guide, particular guru (dead or alive), or by special dispensation from God. In Europe from the 15th to 19th century, doctors and occultists such as Paracelsus and Fludd, and healers such as Mesmer theorised about the likely magnetic nature of energies instead of believing that they were tapping into 'powers' belonging to others.

Touching as a part of the healing act was widely used by both shamans and traditional practitioners until the rise of Puritan culture in Western Europe during the 1600s. At that time too, there was a shift away from primitive healing to modern scientific medicine (Jahnke 1985, Baldwin 1986). Puritan culture equated touch with sex, which was associated with original sin. Religious and scientific views which rejected traditional practices gave succour to a movement which sought to exterminate those who persisted in using them. Millions of people, especially women, acting as healers were caught up in general persecutions and exterminations of those involved in practices which were little understood and often drawn within the same net as witchcraft (Achterberg 1990). Because of the rise of scientific medicine and the strong puritanical ethic, all unnecessary touch was discouraged. It was not until the 1950s that touch as a therapeutic intervention began being researched (Dossey 1988).

All cultures have developed their own customs and taboos around touching. It has been suggested that Western European culture is one of the most touch-deprived in the world. Jourard (1964) watched pairs of people engaged in coffee shops in Puerto Rico, Paris, Florida and London. He counted the number of times one person touched another during one hour. In Puerto Rico the average was 180, in Paris 110, but in Florida it was only 2 and London 0.

It is important to note here that not all patients, because of their cultural upbringing, want to be touched or feel comfortable in being touched by all people. Therefore, before nurses initiate any kind of touching with patients it is necessary to take into consideration the social context and cultural differences concerning touch. For example, there may be differences relating to different parts of the body which it is considered acceptable to touch. Touch which is acceptable between two women may be inappropriate when offered between men and women. Similarly, levels of acceptable touch between men may also differ. Additional dimensions are brought in through religious and other cultural effects. It is usually inappropriate for a Muslim woman to receive hands-on nursing from a male nurse, and so on.

At the same time, everyone needs touch. Infants cannot survive without it, and studies show that the need for touch does not diminish as one grows older. Indeed, the elderly may suffer most acutely from lack of touch because they may have fewer family and friends and are affected by various social stigmas. Touch also functions as an effective communication channel for the older person at the time of life when other forms of communication are less acute (Baldwin 1986).

Touch is probably one of our most highly used senses. The skin is the largest sense organ of the body, and touch is the first sense to develop in the human embryo, the one most vital for survival. Much of the infant's first information about the world is gained from the way he or she is touched during the birthing process, and touch continues to be used to learn about our environment. It is through the skin that knowledge about the external world is communicated to the brain, and in turn people convey to others information about themselves. A piece of skin the size of a £1 coin contains more than 3 million cells, 12 feet of nerves, 100 sweat glands, 50 nerve endings and 3 feet of blood vessels. It is estimated that there are about 50 receptors over 100 square centimetres, a total of about 900 000 sensory receptors per person (Montague & Matson 1979).

Touching has always been an integral part of nursing and many related disciplines, but it is the way we touch that determines whether it will be an act of healing or a mechanistic procedural act. In order for touch to be used as healing, nurses can learn the necessary techniques. Most nurses intuitively move forward to hold a hand or to touch an arm as a gesture of reassurance. Hand holding has been described as a positive means of communication and one that seems to break down barriers (Knable 1981). Through the mechanism of touch, a nurse can convey feelings of caring and understanding to the client (Schmahl 1964). Other feelings, intentionally or unintentionally, can also be conveyed. For example, judgemental attitudes are revealed in the way nurses do or do not touch patients. It is very important that nurses first examine their own feelings about the meaning of touch before using it as a therapeutic tool.

In order for any kind of touch to be truly effective it must be authentically given by a warm, genuine, caring individual to one who is willing to receive it. Touch cannot and should not be packaged and dispensed. Insincere or inappropriate touching may be more upsetting to the patient than none at all.

Recent thinking on the nature of energy

Quantum physics is just one of the new disciplines that is beginning to explore some of the complex energy relationships between people and their environment. There is a paradigm (a map or blueprint of reality) shift from the older mechanical world view of the Newtonian pragmatists to the new perspective of an interconnected holistic universe as envisioned by post-Einsteinian thinkers. Newtonian physics is mechanistic and reductionist – it views the person as essentially a group of body systems interacting in predictable ways. Post-Einsteinian thinking encourages a more holistic view of people – not just an interrelationship of body organs, but connected with the energy of the whole universe. Modern physicists see the universe as a dynamic web of interrelated events, none of which functions in isolation (Capra 1976).

Bohm (1973) wrote of quantum theory as follows:

... there has been too little emphasis on what is, in our view, the most
fundamentally different new feature of all, i.e. the intimate
interconnectedness of different systems that are not in spatial contact, and
later ... the parts are seen to be in immediate connection, in which their
dynamical relationships depend, in an irreducible way, on the state of the
whole system (and indeed on that broader system in which they are
contained, extending ultimately and in principle to the entire universe).
Thus one is led to a new notion of unbroken wholeness which denies the
classical idea of analysability of the world into separately and independently
existent parts.

In simpler language, this new physics believes that energy and mass are the same thing, every living thing in the universe is a pattern of moving energy, and that all living beings are interconnected to all other living things and interacting with them all the time. These ideas require nurses and many other health care workers to move from the concrete to the abstract because they challenge almost everything they have been taught about the nature of the universe and reality.

Our belief systems can determine our view of reality. For example, some beliefs about the world see it as a planet orbiting the sun, others as riding on the back of a turtle (native American Indians), others as flat. Similarly, traditional thinking on the nature of the body as taught to nurses relies heavily on the solid and predictable behaviour of atoms, molecules and cells. A quantum physics view reveals a very different universe of enormous interchange of energy – the body not so much a solid fixed mass, but in constant exchange of energy internally and externally.

Concepts of human energy

Concepts of essential life energy go by a variety of names in different disciplines and different parts of the world. However, scholars who have made a study of these energies have found strong correlations among all of them. Three major perspectives on vital energies that have significantly influenced recent Western thought are Indian, Chinese and Egyptian. Krieger (1973) chose the Indian philosophy as a framework in which to do her pioneering work with TT.

From the perspective of this philosophy, TT is primarily concerned with the use of prana, which is difficult to translate and define in Western culture, but may be seen simply as the vigour and vitality of the body and all the underlying processes, such as growth, breathing and healing. In a healthy individual, prana is abundant, while a lack of prana is the corollorary of disease. Chakras are also part of this framework. They can be thought of in Western terms as energy transformers which take in subtle energy and distribute it to the major glands, nerve centres and organs of the body. Chakras are considered the centre of consciousness in this philosophy instead of the brain.

Some cutting edge scientists in the Western world are moving away from the Newtonian understanding of life and illness. Instead of taking the human being apart to learn about them and why they become ill, identifying the component parts and how they operate or fail, and then how to fix them, they are moving towards the Einsteinian concept. Einstein was searching for a unified field that would embrace all time and space. The post-Einsteinian viewpoint of vibrational medicine· sees the human being as a multidimensional organism made up of physical/cellular systems in dynamic interplay with complex regulatory energy fields. Vibrational medicine attempts to heal illness by manipulating these subtle-energy fields by directing energy into the body instead of manipulating the cells through drugs or surgery. Therefore these cutting edge scientists think of consciousness as being present in all the cells. For example, each cell is seen as part of a 'river of consciousness' by Chopra (1989), a practising endocrinologist and former chief of staff of a major Massachusetts hospital.

Talbot (1991), in his book *The Holographic Universe*, suggests that consciousness is in the energy field of the person. Martha Rogers and her students (Malinski, Barnett, Parsey, Newman, Quinn, Cowling) are the cutting edge nurse theorists whose ideas relate more to this post-Einsteinian concept, and it is these theorists that this book will explore (while not discounting those who place TT in an Eastern philosophical framework).

The emergence of Therapeutic Touch (TT)

Dr Dolores Krieger, then Professor of Nursing at New York University became intrigued as she watched her friend Dora Kunz, then President of the US Theosophical Society, practise laying on of hands. Krieger went on to learn laying on of hands herself, practised it and researched it before beginning to teach it to her Master's degree students in a programme called 'Frontiers in Nursing' in the early 1970s. It was at this time that she coined the name Therapeutic Touch (TT) for this particular method of laying on of hands. Capital letters are used for the first letter of each word to distinguish it from other forms of therapeutic touch, which includes any kind of touching intervention such as massage, stroking, or hand holding. Krieger's students went on to do good quality research on TT (see Ch. 2) and today it is mainstream nursing practice in the United States, since these students and researchers teach in major nursing schools and universities, and others have applied it in practice.

TT is now viewed as a natural human potential which can be learned and practised by anyone who has the 'intention' to help or heal. This intention is at the core of the TT process and is described as an attempt to focus completely on the wellbeing of the patient in an act of unconditional love and compassion (Quinn 1993). For this reason TT has been called a 'healing meditation' (Krieger et al 1979). More detailed discussion of these aspects of TT takes place in Chapter 2.

In the beginning Krieger and others took their framework for supporting TT from Eastern philosophic thought (indeed similarities between Eastern and Western mysticism and quantum physics are noted by many mystics and scientists). However in the past 20 years, research has become more sophisticated as the branch of physics known as quantum mechanics has expanded. Along with this, scholars of Rogers' ideas have made her theory clearer and the framework for, and definitions of, TT have evolved closely with it. Thus two distinct frameworks in the development of TT are emerging. The first is that taken by Krieger and colleagues based on Eastern philosophy, the second is based on Rogers and quantum physics work, and supported strongly by Quinn and other Rogerian scholars in their research (Society of Rogerian Scholars 1993).

Krieger's (1973) first definition of TT is 'a derivative of laying on of hands that uses the hands to direct excess body energies from a person in the role of healer to another for the purpose of helping or healing that individual'. Two years later she added 'Although derived from the laying on of hands, it differs from it in that TT is not performed within a religious context. The person in the role of healer does the act of TT while in a meditative state and is motivated by an interest in the needs of the patient' (Krieger 1975).

Another definition comes only recently from the Nurse Healer's Professional Association (NHPA 1992) an association of nurses in the US who practise TT as a contemporary interpretation of several ancient healing practices, (TT) 'is a consciously directed process of energy exchange during which the practitioner uses the hands as a focus to facilitate healing'.

The idea of energy exchange (or transfer) is often seen as a linear process from practitioner to client. The implied causality, and the separation of the practitioner and client into two interacting human fields is too simplistic for many teachers and practitioners, and some nurse theorists. There appears to be strength in the argument, as will be discussed in subsequent chapters, that the relationship between the nurse giving, and the patient receiving TT, is much more intricate and complex than simple cause and effect.

Some nurses have expressed concerns that complementary or alternative therapies, or technique based on Eastern philosophical thought, 'leads to the destruction of the soul and spirit in Hell'. The same author claims they see all complementary therapies as based on anti-Christian religions that 'may lead to occult involvement' (Gennis 1992). There are concerns that the practice of TT is a 'heathen' activity, based on fear of the person opening up to an outside force. They see the allowing of the mind to go blank (as sometimes suggested in meditation practices) as an invitation to invasion by demonic spirits (Malinski 1993).

Furthermore, because TT is still difficult to demonstrate (other than through experiential exercises) or energy transfer able to be measured in a satisfactory way, it has even been rejected as a hoax. This text will explore these issues in the detailed discussions of many aspects of TT in subsequent chapters.

As the awareness and knowledge around the understanding of TT continues to expand, many nurse theorists and teachers believe it is now time to bring TT into a purely nursing framework. (This is not to discount the extremely valuable work that has been done up to this time.) Rogers' model, The Science of Unitary Human Beings, is the framework that has provided the foundation for most of the TT studies that followed Krieger's initial work. The authors agree that Rogers' model is the conceptual model that we will be using for this book. Rogers defines people as irreducible, indivisible, multi-dimensional energy fields integral with the environmental energy field (Rogers 1990). Energy transfer is too simplistic in relation to Rogers' model. Her theory is so crucial in understanding (and therefore practising) TT that an entire chapter has been included on it (see Ch. 3). It is important to note that while there is increasing support for the Rogerian premise of a human energy field, to date there have been no accurate reports of investigations, within or outside of nursing, which measure interaction between such individual human fields. 'The reality of such interaction therefore remains axiomatic' (Quinn 1989).

'Healing' is another word that is uncomfortable for many. Quinn (1992) suggests that:

Healing, the emergence of right relationship at, between and among all the levels of human being, is always accomplished by the one healing. No one and no thing can heal another human being (but themselves). All healing is creative emergence, new birth, the manifestation of the powerful inner longing, at every level, to be whole.

She believes that the role of the nurse is to remove barriers to the healing process. Florence Nightingale was perhaps one of the first who understood this concept. She said (1869) that the role of the nurse was to put the patient in the best possible condition for nature to act. More recently it has been postulated that the TT practitioner knowingly participates in the mutual human/environment process by shifting consciousness into a state that may be thought of as a 'healing meditation'. This facilitates repatterning of the patient's energy field through a process of resonance, rather than 'energy exchange or transfer' (Krieger et al 1979, Cowling 1990, and Quinn 1992).

The definitions in the literature now begin to change. Meehan (1993) a student of Krieger and a researcher, teacher and practitioner of TT herself, redefined TT as 'a knowledgeable and purposive patterning of patient-environmental energy field process in which the nurse assumes a meditative form of awareness and uses her/his hands as a focus for the patterning of the mutual patient-environment energy field process'.

Malinski (1993) retains some of Meehan's wording when she defines TT as 'a health patterning modality whereby nurse and client participate knowingly in the changing human-environmental field process'.

Quinn clearly brings TT into a more accurate Rogerian perspective when she discusses the interconnectedness of all our lives. She sees our consciousness as not separate and apart but integral with all consciousness. With the intentional use of our own consciousness, we can move towards the repatterning and healing of ourselves and others. She believes that it is possible in the act of 'centering' (see Ch. 4) to shift one's consciousness. This in turn may provide a template of sorts upon which the patient may repattern and balance their energy field, which permits healing to take place.

Using the metaphor of sound, the pattern, or vibration of the nurse's consciousness becomes a tuning fork, resonating at a healing frequency, while the client has the opportunity within the mutual person-environment process to tune, to resonate, to that frequency.

Quinn (1992).

This way of looking at TT is also consistent with the modern physicists. Bentov (1977), a biochemical engineer, appears to share this view when he writes:

the real reality . . . the microreality, that which underlies our solid reality . . . is a rapidly pulsating matrix of fields of energy, an interference pattern of waves filling the vast vacuum of our bodies and continuing beyond them in a more diluted fashion . . . we may look at a disease as an out-of-tune behaviour . . . when a strong harmonising rhythm is applied to it, the interference pattern of waves, may start beating in tune again.

Research to date has suggested that laying on of hands can increase the rate of wound healing in mice (Grad et al 1961), the rate of growth in plants (Grad 1963), the rate of activity of the enzyme trypsin (Smith 1972) and the level of human haemoglobin (Krieger 1972). Outcome studies to date indicate that it can increase human haemoglobin level (Krieger 1974), induce physiologic relaxation (Krieger et al 1979), decrease anxiety (Heidt 1981, Quinn 1989), relieve pain (Keller & Bzdek 1986, Meehan 1985), lower diastolic blood pressure (Quinn 1989), reduce stress in hospitalised children (Kramer 1990) and accelerate wound healing (Wirth 1993). Detailed discussion of this research is included in Chapter 2.

This research cannot be ignored and, of course, much more yet needs to be done. As nurses continue to sift through the rapidly accumulating scientific data at the frontiers of practice, many also find themselves drawn to TT for a different reason . . . simply, that it seems to work.

ONE NURSE'S STORY – HOW I CAME TO T T

It was 0400 as I sat down with a hot beverage at the nurses' station to start my charting. My eyes drifted down the long, dark hall at the Cancer Research Institute at the University of California and stopped at the door out of which I had come five minutes before. I tried to form the words in my mind that I would put in the chart of Joan, the 33-year-old woman whom I had known as a patient with leukaemia for over a year. Joan had experienced two remissions but was now in the hospital with a liver abscess. She had been admitted almost 72 hours before, on my shift and, as in her admissions before, I was her primary nurse. I identified with Joan in her role as a mother and wife, and greatly admired her courage in the way she approached her physical disease with her use of meditation, yoga and other 'alternative' approaches in her quest towards healing. She had been a wise teacher, not only to me but to many other world expert oncologists. A good deal of my time had been spent with her for the last 8 hours (we worked 12-hour shifts), for in spite of every known trick or technique or medication known to modern science available in a University Teaching Hospital, Joan was still not only febrile (up to 40°C), but was in intense pain – tossing, turning, moaning on her bed, unable to fall into sleep. While searching my repertoire for some way in which I could relieve her pain and give her some comfort, I remembered a class in Therapeutic Touch I had taken a few months earlier. At the time I hadn't felt I could integrate the technique into my high-tech nursing. It had been interesting but I felt I couldn't do Therapeutic Touch as well as the others in the class and also was not sure how it matched with my concept of nursing. To tell the truth, I was a little in awe or afraid of the whole idea. But Dee's [Dolores Krieger's] words came back to me: 'Anyone can do TT, anyone can use TT to increase comfort and decrease anxiety'. Joan was beyond judging me, so I gave it a go. I did TT with Joan for the next 10 minutes. She had fallen asleep and here I was now wondering how to put into words the complexities of the whole interaction. I put her chart aside and did routine charting on my other patients. And then, a half hour later, Joan's light went on. What could I possibly do for her that I hadn't already done, I thought as I walked the distance to her room. 'Could you change my wet bedding?' Joan said sleepily as I walked in. Joan had been sweating profusely, her temperature had broken and was now 38°C and her pain was completely gone. I was 'blown away'. The next day my clinical supervisor and I went through all her notes to find some other explanation – not only for Joan's pain relief but for the normalisation of her temperature. We did not find it but, over the rest of Joan's hospitalisation, TT administered either by me or her husband to whom I taught TT, was used to keep her free of pain and her temperature acceptable (along with the appropriate medications which had not been working before).

I know there could be many explanations for what occurred, but none seem to fit with what happened that night. All I know is, it seemed to work. I was able to help someone in need and be confident that I could do no harm in the process.

Why us?

Caring is a central concept underpinning the work of many health care professionals. Nurses, for example, make this the principal focus of their activity. One way of expressing caring is through what Benner (1984) has called 'presencing' – being with the patient in such a way that the patient feels not only cared for, in terms of the practical things that are done to help, but also cared about and valued as a person. This means, for example, that nurses and others do not necessarily always have to like all their patients, but that they can still recognise their value as human beings and seek genuinely and sincerely to help them. Campbell (1984) has described this as form of 'moderated love', where the relationship is not physical or sexual, but filled with the intense desire to help and heal another person.

Through caring, nurses and others are not only able to help sick people recover, they can also help them to feel better in themselves. Even with a patient who may be dying, where getting better is not possible, carers can help by promoting comfort, relaxation, a sense of wellbeing and acceptance.

In addition, this feeling better side of nursing underpins the possibilities of getting better. People who are sick are far more likely to heal, and heal more quickly, when they are helped to feel better about themselves (Kitson 1988). As the following chapters will explore, TT is underpinned by the notions of intentionality (the focused desire on the part of the healer to help the patient) and helping people to feel better. This puts the patients in a better position to mobilise their own healing energy more effectively.

Thus when people feel better, they are far more likely to find, for example, improvements in wound-healing, fewer risks of post-operative complications and so on. In this sense, it is possible to construct a strong case for the cost-effectiveness of TT. If nurses and other health care workers can bring TT into their everyday practice, there are several knock-on effects. Most immediately this concerns the patients, improving their general level of wellbeing and putting them, as Nightingale (1869) noted, 'in the best position for nature to act'. Perhaps one of the challenges for future research into TT is to demonstrate its cost-effectiveness. Time spent helping people to feel better may have cost-benefits to the healthcare system if people get better more quickly as a result.

Another factor is the effect upon the staff themselves. As will be discussed in greater detail in Chapter 4, the relationship of the energy fields between practitioner and patient is neither linear nor unidirectional. Care staff derive benefit from TT as well – for example, in terms of general wellbeing and collegiate relationships. There are implications here not only for patient care, but also for the health of the organisation, staff relationships and effectiveness.

Health Services tend to be the biggest single employers in many societies. They consume huge amounts of resources, occupy thousands of people and affect millions of others. There are over 5.5 million nurses in Western Europe alone. Imagine the implications if every one of them and their millions of colleagues in other disciplines was empowered with the healing potential of TT! Such power would not only affect the patient–practitioner relationship, not just the health-care system, but could help transform society as a whole.

REFERENCES

Achterberg J 1990 Woman as healer. Shambhala, Boston
Baldwin L 1986 The therapeutic use of touch with the elderly. Physical and Occupational Therapy in Geriatrics 4: 45–50

Bentov I 1977 Stalking the wild pendulum. Dutton, New York

Benner P 1984 From novice to expert. Addison Wesley, New York

Bohm D 1973 Quantum theory as an indication of a new order in physics: implicate and explicate order in physical law. Foundation of Physics 3: 139–168

Campbell A 1984 Moderated love. SPCK, Edinburgh

Capra F 1976 The tao of physics. Bantam, New York

Chopra D 1989 Quantum healing. Bantam, New York

Cowling R W 1990 A template for unitary pattern-based nursing practice. In: Barret EAM (ed) Visions of Rogers' science-based nursing. National League for Nursing, New York

Dossey B 1988 Holistic nursing: a handbook for practice. Aspen, Gaithersburg, Maryland

Gennis F 1992 Alternative roads to hell? Nursing Standard 6 (44): 42–43

Grad B, 1963 A telekinetic effect on plant growth. International Journal of Parapsychology 5: 117–133

Grad B, Cadore R J, Paul G I 1961 An unorthodox method of treatment of wound healing in mice. International Journal of Parapsychology 3: 5–24

Heidt P 1981 Effect of Therapeutic Touch on anxiety level of hospitalised patients. Nursing Research 30: 32–37

Jahnke R 1985 The body therapies. Journal of Holistic Nursing. Spring 7–14

Jourard J 1964 The transparent self: self disclosure and wellbeing. Van Nostrand, Princetown

Keller E, Bzdek V M 1986 Effects of Therapeutic Touch on tension headache pain. Nursing Research, 35(2): 101–106

Kitson A 1988 On the concept of nursing care. In: Fairbairn G, Fairbairn S (eds) Ethical issues in caring. Gower, Aldershot

Knable J 1981 Handholding: One means of transcending barriers of communication. Heart Lung 10: 1106

Kramer N 1990 Comparison of Therapeutic Touch and casual touch in stress reduction of hospitalised children. Paediatric Nursing 16(5): 483–485

Krieger D 1972 The response of in-vivo human haemoglobin to an active healing therapy by direct laying on of hands. Human Dimensions 1: 12–15.

Krieger D 1973 The relationship of touch, with intent to help or heal, to subjects' in-vivo haemoglobin values: a study in personalised interactions. (Proceedings, American Nurses Association 9th Nursing Research Conference, San Antonio TX. March 21–23) 39–78.

Krieger D 1974 Healing by the laying on of hands as a facilitator of bioenergetic change: the response of in-vivo human haemoglobin. Psychoenergetic Systems 1: 121–129

Krieger D 1975 Therapeutic Touch: the imprimatur of nursing. American Journal of Nursing 75: 784–787

Krieger D, Peper E, Ancoli S 1979 Physiologic indices of Therapeutic Touch. American Journal of Nursing 14: 660–662

MacManaway B Turcan J 1983 Healing. Thorsons, Wellingborough

Malinski V 1993 Therapeutic Touch: the view from Rogerian nursing science. Visions: The Journal of Rogerian Nursing Science 1 (1): 45–54

Meehan T C 1985 The effects of Therapeutic Touch on experience of acute pain in postoperative patients. Dissertation abstract International 46 795B (University Microfilm no. 8510765)

Meehan T C 1993 Cited in Malinski (1993) op cit.

Montague A, Matson F 1979 The Human Connection. McGraw Hill, New York

Nightingale F 1869 (1980 ed) Notes on nursing, what it is and what it is not. Churchill Livingstone, Edinburgh

Nurse Healers Professional Association 1992 Summer Newsletter New York

Peper E, Ancoli S 1976 Two endpoints of an EEG continuum of meditation. In: Krieger 1979 Therapeutic Touch Prentice Hall, London

Quinn J F 1984 Therapeutic Touch as energy exchange: replication and extension. Nursing Science Quarterly 2(2): 79–87

Quinn J F 1992 Holding sacred space: the nurse as healing environment. Holistic Nursing Practice 6(4): 26–35

Quinn J F 1993 Psychoimmunologic effects of Therapeutic Touch on practitioners and recently bereaved recipients: a pilot study. Advanced Nursing Science 15(4): 13–26

Rogers M 1990 Nursing: science of unitary, irreducible, human beings: update 1990. In: Barrett E (ed) Visions of Rogers' science based nursing. New York, National League for Nursing

Schmahl C 1964 Ritualism in nursing practice. Nursing Forum 11: 74

Smith J 1972 Paranormal effects on enzyme activity. Human Dimensions 1: 12–15

Society of Rogerian Scholars 1993 Visions. Journal of Rogerian Nursing Science, Pensacola

Talbot M 1991 The Holographic Universe. Harper Collins, New York

Wirth D 1993 full thickness dermal wounds treated with non-contact Therapeutic Touch: a replication and extension. Complementary Therapies in Medicine 1(3): 127–132

2

A *review* of the
literature and research

Michael Richardson

■ *There is the danger that the phenomenon may, like the butterfly that is pinned down for closer inspection, be destroyed in the attempts to understand it.*

DR JANET QUINN

What is Therapeutic Touch?

The earliest reference in the literature to Therapeutic Touch (TT) states:

> *Therapeutic Touch . . . consists of the simple placing of the hands for about 10 to 15 minutes on or close to the body of an ill person, by someone who intends to help or to heal that person.*

(Krieger 1975)

It implies that a process occurs, whereby the nurse (the healer) directs attention using the hands to achieve a therapeutic effect for the patient (the healee).

Krieger (1975) added that the practice of TT is a 'natural potential in physically healthy persons who are strongly motivated to help ill people'; most definitions of TT agree with this account (Meehan 1990a, Quinn 1992, Wright 1987).

Later definitions by Krieger and other writers put emphasis on describing it as an exercise in which the healer, intentionally through a process of meditation, acts as a channel for an assumed 'universal life energy' which actualises a healing process in the healee (Krieger et al 1979, Macrae 1988). For this reason TT has also been described as a 'healing meditation' (Peper & Ancoli 1976), and the underlying dynamic by which it has an effect is an exchange of this assumed energy (Quinn 1984).

A basic assumption of TT is that the interplay of energy accelerates a natural self-healing mechanism in the client, as the following description implies:

> *The transfer of energy from the person playing the role of healer is most usually little more than a booster until the patient's own recuperative system takes over. At best the healer accelerates the healing process.*

(Krieger 1979)

It has been postulated that physical contact is not needed to perform TT because the treatment is based on the assumption that all living things are surrounded by an energy field which can be balanced using the hands (Carruthers 1992). The nature of this theorised energy field has not been empirically defined as yet (Dossey 1989, cited by Olson et al 1992). Fedoruk (1985) therefore makes a helpful differentiation between two forms of administering the treatment: Therapeutic Touch *a*, where actual physical contact takes place between the practitioner and the recipient, and Therapeutic Touch *b*, where it does not.

- Many researchers have studied 'touch' in nursing (e.g. McCorkle 1974, Weiss 1986) and one example of the many general definitions is 'touch is the tactile stimulation of one person by another that is either planned or incidental' (Mulaik et al 1991). The technique of TT as introduced by Krieger however involves touching with the intent to help or heal by meditation (Krieger 1975, 1979) and is therefore not entirely synonymous with most general definitions of touch.

TT has been derived from the ancient practice of 'the laying on of hands' (Krieger 1975, Sherman 1985, Quinn 1989b) and Payne (1989) has clarified how it differs from this in five ways:

- the healer does not have to have religious beliefs

- the client does not have to have faith in TT

- it is knowledge based on research

- as a nursing intervention it is based on a specific assessment

- the afflicted individual has the potential to perform TT on himself or herself.

Green (1986) writes that there is a whole spectrum of healing therapies: touch and massage (e.g. Harrison 1986), TT, hand healing (done by people who believe that they have healing powers) and faith healing. The differences between the therapies are based on the belief systems of the healers and the clients. He gives as a minimal description of a healing ritual:

... a person holding and accepting another, both highly aware and attentive, awaiting some change.

Finally, in searching for a definition of Therapeutic Touch, Quinn (1989b, 1992) reminds the reader that the fundamental focus is on healing and 'wholeness' rather than on disease or symptoms of disease. The word heal derives from the Anglo-Saxon word *hælan* meaning *to become whole*, or *harmony of body-mind-spirit*. Quinn, finding the term to heal a definitional challenge, explains it by defining the opposite:

> *When we are alienated, isolated, estranged, fragmented, groundless, or rootless (from our bodies, our deepest self, culture etc.) we are not whole; we are in a wrong relationship; we are diseased . . . the elicitation of the haelan effect, the total organismic response towards wholeness, is the goal of Therapeutic Touch.*

(Quinn 1992)

To summarise the main points so far, the treatment of TT has been described by most researchers and practitioners as a derivative of the laying on of hands, but a process that takes place outside a religious context. It is viewed as a natural human potential that can be accomplished by anyone who wishes to help or heal.

Similar terms such as faith-healing and hand-healing have been mentioned and although not synonymous with the term may have had some similarities in their origins. The next section therefore attempts to build on the previous chapter and locate in the literature the historical development of TT from its ancient beginnings to the work of Krieger and successive researchers.

Ancient beginnings in religion

The earliest figure known for having healing powers by the laying on of hands was Imhotep, a court architect of King Zozer of the Third Egyptian Dynasty, about 2 700 BC (MacManaway & Turcan 1983, Hodgkinson 1990). MacManaway & Turcan described how Imhotep was deified after his death and many temples were erected and dedicated to him where the sick were brought for healing. This healing cult spread via Persia to ancient Greece where the god assumed the name of Asklepios. Asklepios was a Greek physician and spiritual healer who was also deified after his death. It is probable that the two cults are linked since Imhotep was thought to inhabit the body of a snake (MacManaway & Turcan) and Asklepios (the patron saint of all physicians) has as his emblem a snake coiled round a rod, which is still the symbol of medicine (Hodgkinson 1990).

The term 'laying on of hands' for healing purposes is most commonly ascribed to the teachings and life of Jesus Christ (Zefron 1975). However, according to MacManaway & Turcan what Jesus practised was not intrinsically new. They cite many authorities who argue that Mary and Joseph were Essenes (a sect who, according to the Roman historian Flavius Josephus, practised various forms of spiritual healing) and that Jesus received his early education in the *Therapeutae* temples in the years after fleeing from Herod. Chapter 1 shows how TT is viewed as a natural human potential. Zefron refers to passages in the Bible which show how Jesus commissioned his followers and regarded it as not out of the ordinary that they should be able to heal the sick (see for example Mark 9: 14–29, Luke 9: 37–41, and Matthew 17: 14–21).

Hodgkinson describes how in the Western world the concept of spiritual healing had 'virtually died' by the 4th century because there was a deeply held fear that occult practices such as witchcraft could be associated with healing, especially the type which seem to run counter to normal scientific rationale.

For example:

> Until as late as 1951, I and others like me ran the theoretical risk of arrest under the Witchcraft Act which carried the death penalty, even if it had not been used for some time.

(MacManaway & Turcan 1983)

In describing the history of TT, Turton (1988) cites one of the first Catholics involved in the recent 'charismatic renewal' of the Christian Church (MacNutt 1979), who maintains that there are three sources of healing power.

- the divine power of God

- demonic power

- a natural force of healing, based on love, which some people appear to have.

Turton explains that if this recognition of healing, as a nursing technique, is a natural gift, it can absolve both the practitioner and the client from any guilty fear that they may be meddling in superstitious practices. Only one article in the nursing literature was found to discourage TT specifically on the grounds that it may be a superstitious practice, and for that reason it will be mentioned here.

It was claimed by Miller (1987) that:

If one assumes the energy (the universal energy field) is the Holy Spirit . . ., the healer commits blasphemy by assuming a directive, manipulating role with God himself. If on the other hand . . . the energy is some kind of psychic power, then scripture forbids us to take part in it.

The above statement is valid in that it supports the teachings of the New Testament, however Miller fails to explain clearly the differences between TT and faith healing, which is practised in the Christian world. It is also unclear (due to the lack of literature) whether the opinions expressed are representative of other Christian and non-Christian nurses.

The historic assumption that all matter is surrounded by a universal life energy can be found in Brennan's (1988) review of the history of healing through the human energy field. In the Old Testament there are many accounts of light energy surrounding people, but over the centuries the phenomena have lost their original meanings. For example, Michelangelo's statue of Moses depicts him with two horns on his head, rather than the two beams of light which the word *karnaeem* originally tried to convey. The confusion arose because this word can mean either horn or light in Hebrew. Brennan cites White & Krippner (1977), who listed 97 different cultures that refer to the human energy field with 97 different names. In India, sacred writings refer to this life energy as prana and in China since the third millennium BC it has been named Qi (ch'i) (Krieger 1975, MacManaway & Turcan 1983, Brennan 1988, Turton 1988, Hodgkinson 1990, Talbot 1991).

Finally, traditional Cree Indian healing practices have also been linked with the concept of energy fields and Krieger's (1979) technique of TT (Morse, Young & Swartz 1991). For particularly serious cases of illness an Indian healer performs a ceremony where dead eagle wings are used to intersect the *Eagle Spirit*, a healing energy which surrounds a person.

The historic and religious concepts of TT such as the eagle ceremony may create healing energy in ways that from a fundamental scientific view were not understood by the cultures that practised them. The history of healing has a strong association with religion and spiritual disciplines and can still be seen in cultures today which have not abandoned their traditions. Could it be that the ancient traditions of the world are simply different interpretations of the same phenomenon, whether it be a natural/learned human potential, superstitious practice or merely a placebo effect?

PRE-NURSING RESEARCH

In the early 1960s the Canadian biochemist Bernard Grad conducted double-blind studies on the use of TT on mice and barley seeds, with the help of a renowned healer, Oskar Estabany. Krieger (1975), Clark & Clark (1984) and Turton (1988) all cite these studies and explain how they inspired Dolores Krieger in pioneering research and practice of TT in the nursing profession.

Studies on mice

Grad (1961) argued that if healing was the result of some energy force, rather than the power of suggestion, then it would have measurable effects on plants and animals. In his first study, Grad selected 300 standardised mice and marked their skin in a specific manner. One third of the group were allowed to heal without any intervention and these were used as the control group. Another 100 were treated by Estabany with the laying on of hands, and the remainder were held by medical students who claimed to have no healing abilities. According to Krieger (1975) after 2 weeks healing in Estabany's group had accelerated to a degree that could have happened by chance less than 1 in 1000 times. Turton (1988) also describes how Grad concluded that wound healing was definitely and significantly accelerated in those animals that had been held by the healer. Conversely Clark & Clark (1984), who provide long, detailed accounts and analyses of these experiments describe how only on the 15th and 16th days of study did the investigators find significantly smaller wound sizes for mice treated by the healer. On subsequent days their wound sizes were found to have no significant difference between mice treated by the healer and the control groups, and on completion of the study many mice not in the experimental group were found to have a similar degree of healing.

It must be pointed out here that Clark & Clark's article does not go beyond research completed after 1979. Furthermore their discussion concerns research on the laying on of hands and on psychic healing, both of which are only peripherally related to TT. Neither of the studies mentioned in their article concerns research on human beings, and it is the effect on human beings that has been the sole focus of research on TT since the term was first coined by Krieger in 1974.

Studies on seeds

In Grad's (1963) study on barley seedlings, the seeds were soaked in a saline solution to simulate a 'sick' condition (Krieger 1975). They were then divided into three groups, two control and one experimental. The first control group was watered with tap-water, the second with water from flasks held by disinterested persons, and the third group was watered from flasks held by Estabany. The dependent variables were the number of plants per pot, the average height of plants per pot, and plant yield per pot. Again according to Krieger (1975), the experimental seeds (those watered by Estabany) sprouted more quickly, grew taller, and had more chlorophyll than the seeds in the control groups. Clark & Clark (1984) however reported that data analysis by means of a t test indicated that on only one day did the experimental group contain a significantly greater number of plants.

Grad's study on barley seeds did show some significant results; however, they may have been biased as the healer prepared all pots and seeds prior to start of the experiment. Grad's rationale was simply that he lacked personnel to carry out these tasks (Clark & Clark 1984). This threat to internal validity could have been reduced if Grad had employed another person to prepare the sample.

Studies on enzymes

The biochemist Sister Justa Smith (1972) decided to test whether or not healing could positively affect the enzyme trypsin in solution. As with Grad, Oskar Estabany was used as the healer. Her assumption was that if enzyme failure is the ultimate physical cause of disease (as proposed by biomedical evidence), any therapeutic effect should be detectable at the same level. Her sample consisted of four solutions of trypsin. One was retained in its native state and used as a control, the second was treated by Estabany for 75 minutes by gripping the test-tube in which it was contained, the third was exposed to ultraviolet light and then treated by Estabany, and the fourth exposed to a magnetic field. Ultraviolet light possesses a wavelength which slightly denatures exposed trypsin, reducing its activity (Smith 1972).

It was concluded that the healer and magnetic field exerted similar positive effects on the enzyme, since the three solutions displayed an increase in enzyme activity. Although Smith reported standard deviations and means of the activity levels of the solutions, she did not calculate any statistical tests for the significance of the findings (Clark & Clark 1984). According to Smith, there was no difference between Estabany's effect on the native and the partially denatured enzyme. That is, both samples in response to Estabany's 'healing' showed an increase in enzyme activity. It is worth noting that the active site of an enzyme resides in a relatively small part of the molecule, and that a partially denatured enzyme can still function. It is possible that the increase of enzyme activity could have been attributed to the heat from Estabany's hands. Smith reported three replication studies of the experiment, but very few details with respect to any of the replications are discussed, and it is possible that the initial findings may have been due to chance.

It is unclear if Krieger (1975) was aware of the limitations of the Grad (1961, 1963) and Smith (1972) studies since they were not mentioned in her writings on them. Nevertheless, she described how, in the late 1960s and early 1970s, Oskar Estabany visited the United States for a few weeks each year to set up a temporary healing clinic with a well known observer of hand healing, Dora Kunz (Krieger 1975).

Krieger observed the interactions with Estabany and the patients very closely, and explained how a significant number of them were cured physically, stating that:

Most of them had verified medical histories and had been referred by physicians who, when the patients returned for follow-up examination, confirmed their improvement.

(Krieger 1975)

It was these experiences that made Krieger decide to study TT in detail by conducting her own research within the nursing profession.

In exploring the related history of TT the literature has shown how healing practices were prominent in early societies, as they still are in many indigenous cultures today. Traditionally the concept has been interpreted from a religious perspective. However, the experiments of Grad (1965) and Smith (1972) show the growing interest that modern science has had in TT and has led researchers to question the 'ultimate' cause of disease and health. For example, Smith (1972) pointed out that:

Without denying the initial premise that disease can be traced to a malfunctioning enzyme, the next question logically would be: what causes the enzyme to function improperly? Or what causes it ever again to function properly? When it comes to healing, we really do not know how even an aspirin works.

To try and answer the question of how healing works, and to interpret the phenomenon of TT as a healing practice, nursing literature has developed a theoretical basis which covers aspects of nursing theory and modern science.

Is there a theoretical framework for Therapeutic Touch?

A theory has been defined as an 'intellectual invention designed to describe, explain, predict or prescribe' (Dickoff & James 1968). In other words, *theories are not discovered but invented* to understand phenomena. To date, practitioners, researchers and educators have used an energy field model as the theoretical framework for inter-

preting TT (Quinn 1989a). This theory postulates that the outcome of
energy between healer and recipient can be deduced from the wider
conceptual system proposed by Martha Rogers (1970, 1990) (See Ch. 3
for a more detailed discussion.) Rogers explained that people are
open systems of energy. 'The human field extends beyond the
discernible mass which we perceive as man. The human field and
the environmental field ... are coextensive with the universe'
(Rogers 1970). As noted in Chapter 1, the view that humans and the
environment are inseparable is a fundamental component of ancient
religions and Eastern philosophies.

When relating this assumption to TT, Quinn (1984) describes how
when one person interacts with another there is an integration of
energy fields and they become each other's immediate environment.
If a change occurs in either person, it is considered to be an outcome
of this field interaction. This change can be healing.

According to Miller (1979), this healing occurs through the
interaction of the energy fields of man and environment whereby one
field 'actively works in the direction of change'. A particular human
field 'healer' can have an environment which is an integral part of an
'ill' human field.

TT can be viewed as a purposive patterning of the ill person's
field in which the nurse uses his or her hands as a meditating focus
to promote healing (Meehan 1990a). Recipient's energies are then
patterned so that they are 'readily available to use in creating the
necessary momentum toward health' (Boguslawski 1990), and this is
achieved by the nurse assuming a meditative state of awareness
(Meehan 1990b), termed by Krieger (1979) as 'centring'.

Malinski (1991), in exploring Rogers' Science of Unitary Human
Beings, hypothesises that the concept of spirituality is synonymous
with the principle of integrality. In this view, spirituality as
integrality is the belief in the notion of an underlying unity between
man and the environment. Hover-Kramer (1990) explains further by
saying that 'we are in fact non-local in nature, connected with each other
in mysterious ways as evidenced by the almost universal experiences
of telepathy ... and being influenced by others'. Centring during TT
treatment heightens the experience of this integrality within the
practitioner and the recipient (Malinski 1991).

ENERGY FIELDS

Rogers' (1970, 1990) conceptual system has described energy fields, yet it is important to have some concept of what a *field* is. Boleman (1985) describes it as a space within which an action occurs. In the context of non-traditional forms of energy, the human field may partly consist of something called emotional energy (Wright 1991). Wright uses the example of anger which can often be felt by another, even without the obvious presence of verbal or non-verbal language. Perhaps, therefore, the practitioner's intention to help through centring can be felt by the person receiving Therapeutic Touch?

It is important to note here that there have been no reports which measure the interaction of energy fields to date (Quinn 1989a). However the Rogerian premise of a human energy field integrated with an environmental energy field is well supported in the literature by modern scientific thought. Several mechanisms from physics – the science dealing with properties and interactions of matter and energy – apply to the understanding of Rogers' (1970) concepts, and these are summarised below:

RELATIVITY

Perhaps the best known theory from modern physics is Einstein's Theory of Relativity. Einstein postulated that light has a constant velocity and this led to his famous equation relating mass to energy ($E = mc^2$), the basis of understanding atomic energy. This simple equation has some remarkable consequences when applied to the Rogerian concept, because it sums up the equivalence of energy and mass. In other words, mass and energy are interchangeable – two manifestations of the same reality. The basic substance of the universe according to Einstein is not matter but energy (Hawking 1988, Payne 1989, Laffan 1993).

QUANTUM THEORY

According to the laws of classical physics, if an object is warmed it should emit the same kind of blue-white light that it does when it is extremely hot. However, in reality it turns red. Max Planck discovered this principle in 1900 whilst investigating black-body radiation. This led him to announce his 'radiation law', according to which light was not emitted as a continuous flow as previously thought, but was made up of discrete units or 'quanta' of energy (Gribbon 1989).

This discovery was very significant in that it was the first evidence that the subatomic world behaves according to completely different post-Newtonian rules (Brazier 1992). The theory contradicted classical physics since it suggested that light was made of particles (quanta) instead of waves. Actually, physicists have since discovered that light can manifest as either particles or waves. Thomas Young's experiment from 1803 proved that light was made up of waves by shining a light at a barrier with two slits in it. The light passed through the two slits and interfered with each other creating dark patches. Conversely light can also be seen to be made of particles via Einstein's photoelectric experiment. Considering the same two slit experiment as before, a single quantum of light can go through one slit and land in an area that would be dark if two slits were open. Physicists have wondered how the quantum knows whether the other slit is open or closed and some have even speculated that quanta may possess consciousness (Brazier 1992). There is also evidence that quanta only manifest as particles when they are being observed (Padmanabhan 1992). Talbot (1991) cites the physicist Nick Herbert, a supporter of this interpretation who states:

Humans can never experience the true texture of quantum reality because everything we touch or look at turns to matter.

The implication that all matter is an ambiguous 'soup' of quanta and only appears as normal reality when we are looking at it supports Rogers' theoretical system that humans are 'irreducible' energy fields integrated with each other and their environment.

INTERCONNECTEDNESS

Quantum theory showed how local reality may only be true at a certain level yet, according to Chopra (1990), the physicist John Bell developed this further by explaining that the reality of the whole universe is non-local. Bell's theorem shows by a famous mathematical equation that all objects and events are interconnected with each other and respond to one another's changes of state. This concept can be equated with Rogers' principle of integrality as discussed above.

To date, practitioners, teachers and researchers have used an energy model for explaining the phenomenon, and this has also been deduced from the Rogerian conceptual system of the Unitary Human. The energy model has support from advances in modern physics which show that on the subatomic level the classical laws of physics do not apply and that the universe may well consist of interconnected energy of which life is a part.

Research on Therapeutic Touch

When applying the conceptual system of Rogers' Science of Unitary Human Beings (1990) to research studies, the question being asked is: 'How does the outcome of a study have an effect?' No published studies directly measuring energy exchange have been found in the literature, therefore to apply the theory to current research, the *logical consequences of the theory* have to be scrutinised.

Quinn (1989b) suggests six general assumptions that need to be verified in TT research:

1. There is a human energy field which is primary and not secondary to biochemical physiologic processes

2. In illness there is an imbalance in the energy field (the term *imbalance* should be operationalised)

3. Energy moves between human fields

4. Energy can be 'sent' from one person to another

5. 'Recipient' of this energy alters the field of the recipient

6. If energy can be transferred, it impacts on the health of the recipient.

Therefore in reviewing TT research, this critique will include an enquiry into the logical consequences and conceptual limitations of studies which are applicable to one or more of the assumptions laid down by Quinn (1989b), in addition to looking at methodological issues such as research design.

HAEMOGLOBIN

The work of Dr Dolores Krieger in 1974 was the first attempt to scientifically study the effects of TT using trained registered nurses. Krieger (1975, 1979) used an Eastern philosophical view of health to conceptualise this pioneering research, whereby a healthy person has an overabundance of energy, or prana (see Ch. 1), and a deficit of this causes illness.

Prana can be activated by will and can be transferred to another person if one has the intent to do so. The literature also states that prana is intrinsic in what we would call one oxygen molecule.

(Krieger 1975)

It was postulated that a nurse trained to practise TT could activate and facilitate the flow of this healing energy. Krieger's reasoning for choosing haemoglobin was that she claimed it was one of the most sensitive indicators of oxygen uptake (Krieger 1975). Krieger also reasoned that haemoglobin's similarity to chlorophyll was another reason for its selection. She reported that Grad's studies on barley seeds 'showed an increase in the chlorophyll content of the sample that had been irrigated by Estabany ... and that chlorophyll and haemoglobin have a similar 'stereochemical structure". She further reasoned that the selection of haemoglobin would be appropriate for a dependent variable, as haemoglobin is closely associated with the functioning of several enzymes. This deduction was based on Smith's (1972) findings on enzyme activity discussed earlier. Krieger took a convenience sample of 32 nurses out of 75 who asked to participate. The 32 selected nurses were those who Krieger believed best showed the qualities of a 'humanistic nurse', and they were taught methods of TT either by Krieger or by her colleague Dora Kunz. Each trained nurse was asked to choose one patient to receive TT and a second to receive routine care. The sample therefore comprised of 64 hospitalised patients, half of whom were treated with TT, and half given routine care.

The results, which were supported by Fisher's *t* Test for the comparison of correlated means, showed a statistically significant increase in post-test haemoglobin values in those patients in the experimental group. The hypothesis that the haemoglobin levels of those patients would change significantly from their pre-test values was therefore supported.

Methodological problems

The reasoning behind using haemoglobin as the dependent variable warrants criticism, since haemoglobin is not a measure of oxygen uptake, but a measure of oxygen capacity. It has been suggested that an appropriate measure of oxygen uptake would have been the oxygen saturation of blood (Clark & Clark 1984). Krieger's assumptions based on Grad's (1961, 1963) and Smith's (1972) studies are also questionable since Grad (1963) did not use chlorophyll as a dependent measure in his study and Smith did not report testing for statistical difference between pre- and post-experiment measures of enzyme activity.

Absence of operational definitions and random group assignment further weaken Krieger's (1975) study, although she did report using a *t* test to establish the comparability of haemoglobin levels between groups. The *t* test is a powerful parametric test and is the classic technique for analysing the differences between the means of 2 groups, thus one of the assumptions it has to meet is that there are equal variances between these groups (Brink & Wood 1989). Since Krieger (1975) did not report the manner in which the hospital patients were chosen, the significant results have to be treated with caution, because homogeneity in terms of age, sex and type of illness cannot be assumed.

THE RELAXATION RESPONSE

Further collaborative work by Krieger, Ancoli & Peper (1979) reported the first 'evidence' of a deep relaxation state being present in clients receiving TT. This was verified by objective electronic measurements of dependent physiological variables of three patients and a healer (Krieger) during TT. EEG and galvonic skin response were measured for all participants; electromyographic and electrooculographic responses for the healer were noted; and the patients' ECG and hand temperature were recorded.

The electroencephalographic results of the healer demonstrated a predominant amount of fast synchronous beta activity during the procedures, which was interpreted as indicating a state of deep concentration.

On analysing the data, it can be considered that the rapid synchronous beta activity in Krieger's EEG represents the physiological style of TT that . . . can be considered to be in actuality a healing meditation.

(Krieger et al 1979)

The electroocular records of the healer collected during her administration of TT showed that her eyes had 'no movement' which was interpreted by Krieger et al (1979) as an indicator of her steady concentration during the procedure.

Collected data from the patients indicated that there was a high amount of large amplitude alpha activity while their eyes were open. According to the authors this indicated a deep state of relaxation because 'aside from a study done on Zen Buddhist masters, the alpha state is usually accomplished in the closed eyes state by most people' (Krieger et al 1979).

No other significant results were recorded and it is unclear as to why the healer and patients did not have the same physiological variables measured for comparability. In addition, data and statistical tests were not reported in the text. This challenges the credibility of the study, and the conclusions therefore have to be viewed with caution.

The work of Krieger et al (1979), served as a bridge between TT and the relaxation response for at least six other researchers who studied the effect of TT on either anxiety or stress (Heidt 1981, Quinn 1984, Randolph 1984, Fedoruk 1984, Parkes 1985, Quinn 1989a, Olsen et al 1992).

CONTROLLING THE PLACEBO PHENOMENON

Any procedure that produces an effect in a patient because of its therapeutic intent and not its specific nature is called a placebo

(Liberman 1962). For example 'fake' drugs can cause the adrenal glands to secrete and the stomach's blood flow to increase (Wolf 1950), therefore it appears possible that placebos can produce objective physical changes. As the nature of TT is hypothesised as an energy exchange which lies within the wider concepts of Rogerian science, it would appear pertinent that when subjective symptoms of subjects are being studied the inclusion of a placebo control group (under double-blind conditions) is essential to eliminate the possibility of a placebo effect.

The research design of Heidt's (1981) study, which claimed that hospitalised patients who received TT had significantly lower levels of state anxiety, highlights the above proposal. Heidt logically reasoned that if TT could influence levels of physiological relaxation as in Krieger's et al (1979) study, it could also influence the level of anxiety experienced by a subject. A total of 90 hospitalised cardio-vascular patients, who had volunteered, were utilised in the study and randomised into 3 groups. The experimental group received TT by a practitioner for 5 minutes. Another group, control, received casual touch for 5 minutes, consisting of a nurse taking various pulses from each subject.

The third group received no intervention by touch, but instead had a nurse who sat and talked to the patient for 5 minutes. This was designed as a control for the effect of the presence of another helping person.

Pre- and post-treatment anxiety levels were assessed using the Self-Evaluation Questionnaire (Spielberger et al 1970) which on comparison of pre-test and post-test means demonstrated a statistically significant decrease in State Anxiety in the experimental group. This group also had a statistically significant greater reduction in State Anxiety scores than the other two groups and were obtained from analyses of covariance. This test was a more suitable method than Krieger's (1975) use of multiple tests, since it meant that post-test means could be adjusted to account for pre-test differences of State Anxiety.

It is questionable as to whether the research design of Heidt's (1981) study allowed for TT to be interpreted as a major factor in the reduction of anxiety, since there was no control for the placebo effect.

Subjects had received different types of behavioural stimulation and had known when TT was being administered to them. The hypothesised energy transfer in this case therefore could not be tested.

MIMIC THERAPEUTIC TOUCH

The first research which exposed two groups to what appeared to be the same treatment was Randolph's (1980, 1984) study on the physiological responses of female college students to stressful stimuli whilst receiving TT. Randolph (1984) built on the previous work of Krieger et al (1979), which claimed that TT produced a relaxation response. Randolph (1984) however took this reasoning further by hypothesising that the physiological stress response of persons treated with the imitated TT would exceed the physiological stress response of persons treated with genuine TT. Muscle tension, galvanic skin resistance, and temperature were measured as indicators of state anxiety.

A film entitled *Subincision* served as a stressful stimulus. This was a silent footage film showing a sequence of operations performed on tribal adolescent boys. An experimental group of 30 subjects received TT and a control group of 30 subjects received imitated treatment. Whilst they watched the film they were simultaneously monitored for signs of anxiety (see above).

No significant differences in the post-treatment dependent variables were reported and both groups responded to the film with a recognisably similar amount of stress. This might be viewed as a lack of support for the contextual theory that energy exchange is the means by which TT has an effect, since both groups received the same behavioural stimulus. The study was also the first published investigation that was double blind in design since neither the subject nor the data collector were informed of the subject's group assignment until after the session. Although TT was performed by nurses who had been trained to administer the treatment, the question arose as to how the level of ability that had been achieved by the TT practitioners could be validated. Randolph (1984) suggested that TT research would benefit from the development of a tool to address this problem.

It has been suggested that the main factor that produced the non-significant results was the level of health of the subjects (Randolph 1984). Randolph used healthy college students, whilst previous research (Krieger 1975, Heidt 1981) employed less healthy subjects. As the stress response is a normal physiological process which protects the person from external threats then perhaps TT would be unable to suppress such a response.

This issue arises again in later research by Fedoruk (1984, 1985) who studied the effect of TT and a mimic treatment on the response to stress in neonates in intermediate and intensive care. Rosalie Fedoruk's unpublished doctoral dissertation sought to answer the question of whether Therapeutic Touch *b* (non-contact) would reduce the stress of the nursing procedure of measuring 'vital signs' in premature infants. The outcome measures of infant state and transcutaneous oxygen blood gas pressure ($TcPO_2$ were used to measure stress.

Quinn (1988) cited no rationale for Fedoruk's choice of transcutaneous oxygen blood gas pressure as a dependent variable and therefore it is presumed from the text that higher levels of $TcPO_2$ correspond with greater levels of relaxation and oxygen uptake. Fedoruk's reasoning may also have been influenced by Krieger's (1975) studies on haemoglobin values which postulated that the energy transferred during TT is bound to the oxygen molecule. Infant state was measured using the 'Assessment of Premature Infant Behaviour scale (APIB)', which was developed by Als, Lester, Tronick and Brazelton.

A convenience sample of 17 premature neonates was studied and each infant received each of the interventions twice. Repeated measures analysis of covariance were used to test the hypotheses. These analyses showed that infants treated with TT demonstrated a greater change from a higher more aroused state to a lower more relaxed state during observation time than infants receiving mock TT or no TT.

No statistical differences were found between $TcPO_2$ levels between groups indicating that TT had no effect on the physiological indicators of stress.

Fedoruk warned that the results could not be interpreted with any degree of certainty. The cautions conveyed in Fedoruk's study by Quinn (1988) included:

- the sample size of 17 was considered small
- the treatment times were not equal; Therapeutic Touch lasted for five minutes whereas the control treatments lasted for one minute.

As in Randolph's (1984) study, the issue of whether TT should suppress the appropriate response to external stressors arises again:

When one examines the major assumptions of the study . . . one finds that relaxation is seen as the equivalent of suppressing reactivity to real stressors in the external environment.

(Quinn 1988)

It is probably therefore a healthy and normal response for an infant to react to stimuli in the environment as Randolph's research demonstrated, and it is unlikely that TT would suppress such a response.

MIMIC THERAPEUTIC TOUCH AND SUBJECTIVE STATE ANXIETY

Support for the hypothesis of energy transfer depends on how well-controlled studies are. It has already been shown that there is a need to control for the placebo phenomenon. Indeed it has sometimes been difficult to reflect the results back to the theory, for example Krieger (1979) and Heidt (1981), because the studies have not been derived directly from the Rogerian conceptual system. This theory testing was addressed directly by Quinn (1984) and Parkes (1985) and replicated again by Quinn (1989a).

The studies utilised subjective state anxiety as a measure of the efficacy of TT because the effect of TT on state anxiety had already been researched (Heidt 1981). Mimic TT (originally called *Non-Contact*) was operationally defined for the first time by Quinn (1982) as:

. . . an intervention which mimics the movements of the nurse during TT, but during which there is no attempt to centre, no intention to assist the subject, no attuning to the condition of the subject, and no direction of energy.

TT in these three studies was practised without using physical contact, as it was reasoned that physical contact would not be needed if TT involved a type of energy exchange. This treatment was defined as *non-contact Therapeutic Touch*.

State Anxiety was defined as a transitory emotional state or condition of the human organism that is characterised by subjective, consciously perceived feeling of tension and apprehension and heightened autonomic nervous system activity (Quinn 1984, 1989a). This was measured by a Self-Evaluation Questionnaire, STAI form (State Trait Anxiety Inventory), developed by Spielberger, Gorsuch and Lushene (Spielberger 1970).

In the first study Quinn (1982, 1984) randomly assigned a convenience sample of 60 hospitalised adults with cardiovascular disease to either a group receiving TT or a group receiving the mimic treatment. These mimic interventions were carried out by 3 nurses who had no knowledge of TT. All subjects completed a STAI questionnaire before and after the study. Both treatment lasted for approximately 5 minutes.

The findings of Quinn's (1982) study supported the hypothesis that there would be a greater decrease in post-test state anxiety scores in subjects treated with non-contact TT than mimic TT. This was supported at the 0.0005 level of significance lending confidence to the theorem that TT involved an energy exchange.

However in a doctoral dissertation by Parkes (1985, cited by Quinn 1988) the effect of TT on the state anxiety of gerontological hospitalised patients was studied. It was found that there were no significant differences in reduction of anxiety in the group that received TT from the control groups. This study used a similar research design to Quinn's (1982), except there was an additional control group where the treatment nurse held her hands over the shoulder area of the subjects.

Three groups of 20 hospitalised elderly patients between the ages of 65 and 93 years had completed the STAI questionnaire before and after the study, and the comparison of post-test means using analysis of covariance had revealed no differences that were statistically significant.

Finally Quinn's replication of her 1982 study (Quinn 1989a), like Parkes' (1985) study also failed to support the energy exchange hypothesis. This replication differed from her first study which had supported the conceptual theory by one major factor: the mimic TT treatment was administered without eye contact. Quinn's derived theorem from the Rogerian conceptual system was that eye contact would not be needed for TT to be effective. In fact the significant results of her last study could have been related to some form of subtle communication such as facial cues from the practitioner.

Quinn's replication study (Quinn 1989a) used a sample of 38 women and 115 men who were hospitalised awaiting open-heart surgery on the following day. As in the original study (Quinn 1982), the STAI questionnaire was used to measure state anxiety. The dependent variables of blood pressure and heart rate were also measured on each subject and pre- and post-test measures were tested using separate analyses of covariance. No significant differences were recorded in any of the variables for those subjects who were treated by TT or those who received the mimic treatment.

The negative findings of Quinn's (1989a) and Parkes's (1985) study might thus be viewed as lack of support for the Rogerian conceptual theory by which TT may have an effect. If eye contact is needed, it may be possible that some form of placebo action is in operation, whereby the recipient of TT can 'sense' the practitioner's intent to help them.

However, methodological designs combined with contextual factors also need to be taken into account for the interpretation of results. Unlike Quinn's original study (Quinn 1982), the nurse in the replication study (Quinn 1989a) provided both TT and mimic TT. Meditation and 'intentionality' are intrinsic to the TT process. It may therefore be possible that meditation and the purpose of intent cannot be turned on or off by conscious choice or will by the experienced practitioner. Quinn (1989a) supported this notion by explaining that it is possible to be in two states of consciousness at the same time, for example when driving a car. This may have explained the closeness of results in both groups.

Although the results in Quinn's (1989a) study were not significant, all post-test measures did change in the predicted direction with the largest changes occurring in the TT group. Quinn suggested that premature ending of the treatment may have resulted in failure of the dependent variable changes to reach significance because the treatment had only lasted for 5 minutes.

IS EYE CONTACT NEEDED FOR THERAPEUTIC TOUCH?

The negative findings of Quinn's (1989a) replicate study can be interpreted as suggesting that eye and facial contact are necessary for TT to have an effect and therefore a mechanism other than energy exchange is responsible such as the placebo phenomenon.

However, if further investigations demonstrate that eye and facial contact are important, this would still not rule out that energy exchange is responsible for TT. Brennan (1988) explained that the eyes can be used for a powerful focus of energy and therefore absence of such contact may hinder the healing process.

Although no other research exists in the literature that has investigated facial contact during TT, the hypothesis that the eyes may help facilitate this energy exchange might be supported by one of the very few qualitative studies to investigate TT. An as yet unpublished study by France (1991) used a multiple perspective approach to discover the essence of a child's perception of TT. Phenomenology was the guiding philosophy for the selected methods and design, which included serial videotaped interview sessions, children's drawings, parental journals, and the researcher's journal.

11 children aged 3 to 9 years were given TT. Phenomenological reduction and analysis of the data discovered an emerging theme of unity between the subjects that was described by the researcher as 'that look'. The researcher captured this unity in the metaphor 'the eyes tell all'. It appears that the eyes of the practitioner may have left an impression on the children while they received TT. It could have conveyed the desire or intent to help and may have supported the idea that eyes are a factor in concentrating energy or aiding intentionality. This could be further supported by electroocular records of a healer in Krieger's second published study (Krieger et al 1979) which showed no eye movement, indicative of a high level of concentration in the healer.

DISTANCE HEALING AS ENERGY EXCHANGE

The work of Beutler et al (1988) deserves a mention as one of the few experimental trials on TT to be carried out by non-nurses. The study sought to discover whether the laying on of hands and distance healing by a renowed healer could reduce hypertension.

Distance healing, whereby the healer 'transmits' healing to an individual who may be many miles away, is practised by some TT practitioners in North America and the UK (MacManaway & Turcan, 1983). Within the context of Rogerian science, where humans and environment are considered as integrated energy fields interconnected with each other, it may be postulated that a healer and recipient could exchange energy whilst not being in immediate proximity to each other.

Beutler et al (1988) randomised patients into three treatment groups: healing by the laying on of hands (n=40), healing at a distance (n=40), and no healing (controls, n=38). Healing at a distance and no healing were investigated double blind.

It was reported that diastolic blood pressure fell in all three groups from week to week during the period of the trial, and that this was probably caused by a placebo effect of the trial itself. The researchers concluded that no treatment was consistently better than another, and therefore the data could not be taken as evidence of a paranormal effect on blood pressure.

The phenomenon of distance healing and energy exchange would appear therefore not to be supported. However the data study by Beutler et al (1988) did record that each week diastolic pressure was consistently lower in patients treated by healing at a distance compared with those in the controls. Although paired comparisons yielded no significant differences, both groups were investigated under double blind conditions and therefore the reduction in the predicted direction may possibly have indicated a paranormal influence such as energy exchange.

THE EFFECT OF THERAPEUTIC TOUCH ON PAIN

To date, there have only been two studies which have investigated the effects of TT on the experience of pain and these have presented conflicting results.

Meehan (1990a) examined whether TT was effective in decreasing pain in postoperative patients and Keller & Bzdek (1986) investigated its effects on tension headache. Both researchers built on previous work which had indicated that TT had decreased anxiety (Heidt 1981, Quinn 1984), and reasoned that since anxiety influences the pain experience (Hayward 1975), TT should decrease pain. Meehan (1990a) hypothesised that subjects treated by TT would have a greater decrease in post-test acute pain than subjects treated by mimic TT. She also investigated whether the effects of TT would decrease pain more than standard treatment. Neither hypothesis was supported and it was found that the standard treatment of giving medication was significantly more effective than the administration of TT.

Keller & Bzdek (1986) hypothesised that subjects who received TT would experience greater headache relief than subjects who received mimic TT, and that their pain reduction would still be greater 4 hours after the treatment. This hypothesis was supported by statistically significant results of post-test scores, which dropped an average of 70% in the TT group and 37% in the control group.

Both research designs were rigorous in these two studies, using operational definitions of treatments and time length (5 minutes), pain scale questionnaires and suitable data analysis test. However, Meehan (1990a) used a large sample size of 108 subjects, whereas Keller & Bzdek's (1986) sample size was 60 which may have accounted for the statistical differences. Meehan also explained that a very conservative statistical test for significance was used (multiple regression analysis with post-hoc comparison of means) and that a less rigorous measure would in fact have supported her hypothesis. Like Quinn's (1989) study on anxiety, the post-test measures of pain in this study did change in the predicted direction.

The results of these two pain studies therefore appear to support the hypothesis that pain can be relieved by TT. The research designs which were double blind and included a mimic treatment also support the Rogerian conceptual theory of energy exchange by the logical consequences of results.

LENGTH OF TREATMENT FOR THERAPEUTIC TOUCH

Predetermining the TT treatment time for 5 minutes works very well in terms of operationally defining the intervention for research purposes (Quinn 1989b). It has worked well in producing significant outcomes for Heidt (1981), Quinn (1984) and Keller & Bzdek (1986).

However, Meehan (1990a) reported that in her pain study, 5 minutes may not have been long enough for an analgesic effect. As the decrease in pain in the TT group was very close to being significant, it may be reasonable to assume that a longer treatment time may have produced a greater effect against pain.

The poor results of the replicate study of anxiety (Quinn 1989a) were also attributed partly to this problem:

There were many sessions when the treatment nurse sensed that the treatment was just beginning to have an effect and that if a few more minutes were permitted the subject would indeed have been more relaxed.

(Quinn 1989a)

It is suggested that caution needs to be taken against regarding this as the main reason for failure of Quinn's results. However it can be expected that some individuals would respond to TT more quickly than others and would therefore require longer treatment.

A preliminary study which sought to answer questions about the effectiveness of TT in reducing post-traumatic stress for 23 individuals (Olson et al 1992) addressed this methodological issue. This research was implemented in the 2 months that followed Hurricane Hugo in the United States with a convenience sample of volunteers who had experienced personal stress. This was confirmed by a short questionnaire that addressed the effects of the hurricane on their lives.

A repeated session design was used with data collected before, during, and after each session, with heart rate, blood pressure and skin temperature used as the physiological dependent measures. The psychological dependent variables were state and trait anxiety which were measured by two visual analogue scale questionnaires. State anxiety was defined in the study as how much stress the subject felt at that time, and trait anxiety as how much general stress the subject felt most days of his or her life.

The first two sessions were identical in that TT was administered by an experienced practitioner, who terminated the treatment at varying times according to 'subjective cues' from the client. The third session was a control session where the subject and healer sat quietly for 20 minutes in close proximity to each other.

The findings demonstrated that mean state anxiety scores decreased significantly after the TT sessions and also when compared to the control group. However there was no significant differences in pre- and post-test trait anxiety scores. Physiological outcomes reflected relaxation trends but were not significantly different between treatment and control groups. Repeated measures of covariance were used to test these differences.

The decrease in state anxiety and length of treatment was significantly correlated with session length, which ranged from 6.8 minutes to 20 minutes, which suggest that perhaps future research on TT should allow more time for the healer to administer treatment.

Some of the limitations of the study by Olson et al (1992), however, were that the sample size was relatively small compared to previous research. Lack of randomisation and the small number of subjects that returned for the control sessions further weakened the results.

It is interesting from a contextual viewpoint that the control session in this study was similar to Heidt's (1981), whereby the healer sat by the patient for a specific amount of time. It is unclear why Olson et al (1992) abandoned the mimic treatment as a suitable control. Their reasoning was that their method 'allowed the subject and toucher to be in the same environmental field without activity and avoided the mimic treatment criticised as confounding the results of other studies, for example Quinn (1989a)'. Quinn (1989a) in her replicate anxiety study suggested, however, that persons not experienced in TT are the best candidates for mimic TT practitioners, because they would not drift into a consciousness of 'sending energy'.

A placebo effect may have accounted for the significant results of the investigation by Olson et al (1992), because the subjects received two different forms of behavioural stimulation from the healer. The theory of energy exchange can therefore not be supported from this study.

THE AGE FACTOR AND THE EFFECTIVENESS OF THERAPEUTIC TOUCH

There is evidence from at least two studies in the literature to suggest the older person is less responsive to TT (Quinn 1984, Parkes 1985). Quinn's (1982, 1994) study on anxiety found a low correlation between age and response to TT, such that it appeared that the older the subject the less post-test anxiety decreased. The results of Parkes's (1985) investigation were also consistent with these results. However, Quinn (1988) urged caution in interpreting the results by questioning the appropriateness of the STAI questionnaire for older people. This becomes more evident when one examines Parkes's (1985) study on hospitalised elderly patients where it was reported that many of the subjects were concerned with trying to remember the answers from the pre-test.

A single subject design investigating the effects of TT on the quality of sleep of elderly nursing home residents (Braun et al 1986) found that five out of six of the subjects had a better quality of sleep during nights when TT had been administered to them at bedtime. Quality of sleep was measured using Visser's (1979) Sleep Quality Questionnaire. The findings appeared to confirm that TT was effective in enhancing the sleep quality of elderly people. It was suggested by the authors that this technique be added to nursing measures used to help the elderly.

Rather than compare the effect of TT on groups of subjects, this study used a single subject design, which meant that each individual acted as their own control. One of the advantages of single subject designs is that they can be carried out using a small sample size.

The authors did not describe any limitations to their study, however when examining the problem from an energy exchange viewpoint, the possibility of a placebo response cannot be eliminated, as residents were aware which evenings TT had been administered to them. Perhaps future single subject designs could be double blind, for example in this study residents could have had TT administered to them once they were asleep and could also have been doubly unaware of which nights these were.

With, for example, the exception of the study by France (1991), most of the research reviewed in this chapter has used a quantitative approach to investigating the TT phenomenon. Braun's sleep study however was pioneering, in that it was the first published research to approach the subject from a more qualitative direction. TT has been described as a deeply personal experience for both the nurse and the patient (Heidt 1991). The study by Braun et al highlights the need for more single subject designs and qualitative measures in quantitative designs, in order that the personal experience of TT can be investigated. New questions may emerge from such studies which may stimulate and develop the theory of TT within the Rogerian world view.

THE NEED FOR QUALITATIVE RESEARCH

To date only one other published TT research study has included qualitative measures within its design. The investigation of the effects of TT on post-traumatic stress by Olson et al (1992) included a collection of qualitative data which addressed the question of whether the person administering TT and the person receiving it agreed about the body locations where sensations related to TT were perceived.

A figure form with front and back line drawings of the human body was used after each treatment, on which both subject and healer, independently, marked any places on the figure that corresponded to sensations on the subject's body perceived during the treatment. Krieger (1979) had suggested that the figure form would be a useful tool for students in TT, but there are no records of reliability for this method in the research literature.

The results recorded a mean of only 13% agreement between healer and subject of where the sensations were felt on the subject's body. However, comments about the nature of their sensations ranged from 'nothing felt' to 'tingling' in both subjects and healer. The researchers reported many similarities when such comments between healer and subject were juxtaposed. This may be useful for later research in developing alternative tools to determine the interaction experience between healer and subject during TT. The only published wholly qualitative study on TT was by Heidt in 1990 (although Lionberger had completed an unpublished doctoral dissertation in 1985). Using an inductive approach she analysed nurses' and patients' experiences of TT.

A comparative method of data analysis from interviews and observations of seven nurse TT practitioners and seven subjects found that in many instances the experiences of the patients during the treatment paralleled those of the nurses.

When nurses had the inner experience of themselves as patterns of energy, which by nature are 'wholes' rather than 'parts', the patients reflected that shift in experience as well. Similarly, as the nurses stilled themselves and projected that sense of calm to the patient, the patient's energy stilled and they focused on the inner healing process. This 'transfer of energy' took place on a physical as well as psychological level (Heidt 1990).

The growing number of case studies on TT in the literature also provide a valuable qualitative type of indication of the experiences of patients and nurses. An analysis of these studies is beyond the scope of this literature review, however a complete list (to date) is available in the bibliography.

One such case study, which comparatively reviewed the sessions of two patients who were treated by a TT practitioner (Heidt 1991), concluded that the patient who suffered from state anxiety was more amenable to treatment than the patient who suffered from trait anxiety. This reflected the results of the study by Olson et al (1992) of post-traumatic stress which found similar results for the effects of TT on trait and state anxiety. This demonstrates the importance specific case studies can have in furthering the knowledge-base of the phenomenon.

Conclusion

In examining TT literature and research this review has shown how interest in the treatment has grown since is introduction into nursing in 1974. In searching for definitions of the phenomenon, it has been suggested that it can be regarded as a natural potential which any person could perform. TT is a contemporary interpretation of ancient healing practices, yet remains outside a religious context. The early experiments by Grad (1963) and Smith (1972) influenced Krieger to study the concept further and introduce it to the nursing profession. Nursing theory has attempted to explain the method, using an energy exchange model within the wider framework of the Rogerian world view. Finally, a critique of nursing literature has been presented which highlights a number of issues. The theoretical context behind the studies have remained consistent yet the testing of this theory remains inconclusive. An attempt has been made to apply the theory of energy exchange logically to research by examining the research designs and the consequences of their results. If practitioners and researchers continue to use Roger's Science of Unitary Human Beings (1970, 1991) to describe and explain TT, it is important that more research tests this theory directly as Quinn (1984, 1989) and Parkes (1985) did. The main issue that has arisen from this theory testing is the appropriateness of certain controls for the placebo response since it appears important that subjects receive the same behavioural stimulus. It may also be pertinent to question whether what has been described as the placebo response is itself a re-patterning of a subject's energy field which helps them regain health. This highlights the need for interdisciplinary research to validate the existence of such an energy field.

Out of the 12 experimental studies analysed in this critique, only seven presented any type of positive results. The need for replication and further developments and improvement in research designs will substantiate what so far can only be regarded as preliminary 'proof' that TT can improve a person's health. Qualitative research may address this problem by exploring the phenomenon on a more personal level, yet to date this is lacking.

It is hoped that this literature review has presented the strengths and shortcomings of what is a relatively unresearched subject in nursing. It needs to be pointed out, however, that tentative empirical support for TT should be no reason to discourage the practice. Many people have voiced an increased wellbeing from the treatment, evidenced from the number of case studies on the phenomenon, and there are no reports showing it to be harmful (Haviland 1986).

In 1975, Krieger implied TT was the hallmark of nursing (Krieger 1975), since it is an intrinsic attribute of the profession that almost every nursing skill is performed with the act of touch. It may be argued further that TT is the hallmark of nursing because the essence of the phenomenon is the *intent* of the healer to focus completely on the wellbeing of a patient. This process of unconditional love or positive regard may be the most quintessential act of caring a nurse can perform.

REFERENCES

Beutler J J, Attevelt J T, Schouton S A, Faber J A, Mees E J, Geijskes G G 1988 Paranormal healing and hypertension. British Medical Journal 296: 1491–1494

Boguslawski M 1990 Unitary human field practice modalities. In: Barret E A M (ed) Visions of Rogers' science based nursing. National League for Nursing, New York

Boleman J 1985 An introduction to physics. Prentice Hall, Englewood Cliffs

Braun C, Layton C J, Braun J, 1986 Therapeutic Touch improves residents' sleep. American Health Care Association Journal. 12(1): 48–49

Brazier C 1992 An investigation into the paranormal. New Internationalist. 237: 4–7

Brennan B A 1988 Hands of Light: A guide to healing through the human energy field. Bantam, London

Brink P J, Wood M J 1989 Basic steps in planning nursing research: from question to proposal. Jones and Bartlett, Boston

British Medical Association 1986 Alternative therapy: report of the Board of Science and Education

Carruthers A M 1992 A force to promote bonding and well being. Professional Nurse 7(5): 297–300

Chopra D 1990 Quantum healing. Bantam, London

Clark P E, Clark M J 1984 Therapeutic Touch: is there a scientific basis for the practice? Nursing Research 33(1): 37–41

Dickoff J, James P 1968 A theory of theories: A position paper. Nursing Research 17: 197–203

Dossey L 1989 Recovering the soul: a scientific and spiritual search. Bantam New York

Fedoruk R B 1984 Transfer of the relaxation response (Doctoral Dissertation, University of Maryland). Cited in Quinn J F 1988 Building a body of knowledge. Journal of Holistic Nursing 6(1): 37–45

Fedoruk R B 1985 Transfer of the relaxation response: Therapeutic Touch as a method for reduction of stress in premature neonates. Dissertation Abstracts International 46: 978

France M N E 1991 A phenomenological inquiry on the child's lived experience of perceiving the human energy field using Therapeutic Touch. Dissertation Abstracts International 52(12): 6315–6316

Grad B 1961 An unorthodox method of treatment of wound healing in mice. International Journal of Parapyschology 3(1): 5–24

Grad B 1963 A telekinetic effect on plant growth. International Journal of Parapyschology. 6(6): 473–498

Grad B 1965 Some biological effects of the 'laying on of hands' – a review of experiments with animals and plants. Journal of American Society for Psychical Research. 59(1): 95–127

Green R 1986 Healing and spirituality. The Practitioner 230: 1087–1093

Gribbon J 1989 Quantum rules OK! New Scientist: Inside Science 25 (Septermber 16)

Harrison A 1986 Therapeutic Touch: getting the massage. Nursing Times 82(48): 34–35

Haviland D 1986 Safety of faith healing. The Lancet March 22: 684

Hawkings S J 1988 A brief history of time. Bantam, London

Hayward J 1975 Information: a prescription against pain. RCN, London

Heidt P R 1981 Effects of Therapeutic Touch on anxiety level of hospitalised patients. Nursing Research 30(1): 32–37

Heidt P R 1990 Openness : a qualitative analysis of nurses' and patients' experiences of Therapeutic Touch. Image: Journal of Nursing Scholarship 22(3): 180–186

Heidt P R 1991 Helping patients to rest: clinical studies in Therapeutic Touch. Holistic Nursing Practice 5(4): 57–66

Hodgkinson L 1990 Spiritual Healing. Piatkus, London

Hover-Kramer D 1990 Energy fields: implications for the science of caring. Imprint 37(3): 7–11

Keller E, Bzdek U M 1986 Effects of Therapeutic Touch on tension headache pain. Nursing Research 2(35): 101–105

Krieger D 1975 Therapeutic Touch: the imprimatur of nursing. American Journal of Nursing 5(75): 784–787

Krieger D 1979 The Therapeutic Touch: how to use your hands to help or heal. Prentice Hall, New York

Krieger D 1990 Therapeutic Touch. Imprint 37(3): 86–88

Krieger D, Ancoli S Peper E 1979 Searching for evidence of physiological change. American Journal of Nursing 79(4): 660–663

Laffan G 1993 A new holistic science. Nursing Standard 7(17): 44–45

Liberman R 1962 An analysis of the placebo phenomenon. Journal of Chronic Diseases 15: 761–783

Lionberger H 1985 An interpretive study of nurses' practice of Therapeutic Touch. Doctoral Dissertation, University of California, San Francisco

McCorkle R 1974 Effects of touch on seriously ill patients. Nursing Research 23(2): 123–125

MacManaway B, Turcan J 1983 Healing: the energy that can restore health. Thorsons, London

MacNutt F 1979 Healing. Bantam, London

Macrae J 1988 Therapeutic Touch: a practical guide. Arkana, London

Malinski V M 1991 Spirituality as integrality: a Rogerian perspective on the path of healing. Journal of Holistic Nursing 9(1): 54–64

Meehan T C 1990a Theory development. In: Barrett E A M (Ed) Visions of Rogers' science-based nursing. National League for Nursing, New York

Meehan T C 1990b The science of unitary human beings and theory based practice: Therapeutic Touch. In: Barrett E A M (Ed) Visions of Rogers' science-based nursing. National League for Nursing, New York

Miller A 1987 Should Christian nurses practice Therapeutic Touch? Journal of Christian Nursing 4(4): 15–19

Miller L 1979 An explanation of Therapeutic Touch using the science of unitary man. Nursing Forum 18: 278–287

Morse J M, Young D E, Swartz L 1991 Cree Indian healing practices and Western health care: a comparative analysis. Social Science Medicine 32(12): 1361–1366

Mulaik J S, Megenity J S, Cannon R B, Gilead M P 1991 Patients' perceptions of nurses' use of touch. Western Journal of Nursing Research 13(3): 306–323

Olson M, Sneed N, Bonadonna R, Ratliff J, Dias J 1992 Therapeutic Touch and post-Hurricane Hugo stress. Journal of Holistic Nursing 10(2): 120–136

Padmanabham T 1992 Bridge over the quantum universe. New Scientist 10(37): 25–29

Parkes B 1985 Therapeutic Touch as an intervention to reduce anxiety in elderly, hospitalised patients. Doctoral Dissertation, University of Texas, Austin

Payne M B 1989 The use of Therapeutic Touch with rehabilitation clients. Rehabilitation Nursing 14(2)

Peper E, Ancoli S 1976 Two endpoints of an EEG continuum of meditation. In: Krieger D 1979 The Therapeutic Touch. Prentice Hall, London

Quinn J F 1982 An investigation of the effect of Therapeutic Touch done without physical contact on state anxiety of hospitalised cardiovascular patients. Dissertation Abstracts International 43: 1797b

Quinn J F 1984 Therapeutic Touch as energy exchange: testing the theory. Advances in Nursing Science 6(1): 42–49

Quinn J F 1988 Building a body of knowledge. Journal of Holistic Nursing 6(1): 37–45

Quinn J F 1989a Therapeutic Touch as energy exchange: replication and extension. Nursing Science Quarterly 2(2): 79–87

Quinn J F 1989b Future directions for Therapeutic Touch research. Journal of Holistic Nursing 7(1): 19–25

Quinn J F 1992 The senior's Therapeutic Touch education programme. Holistic Nursing Practice 7(10): 32–37

Randolph G L 1984 Therapeutic Touch and physical touch: physiological response to stressful stimuli. Nursing Research 33 (1): 33–36

Rogers M E 1970 An introduction to the theoretical basis of nursing. F A Davies, Philadelphia

Rogers M E 1990 Science of unitary human beings. In: Barrett E A M (Ed) Visions of Rogers' science-based nursing. National League for Nursing, New York

Sherman J 1985 The laying on of hands. Nursing Times 81(47): 18–19

Smith M J 1972 Paranormal effects on enzyme activity. Human Dimensions 1: 15–19

Spielberger C D 1970 Anxiety and behaviour. New York. Cited in Heidt (1981)
Talbot M 1991 The Holographic Universe. New York, Harper
Turton P 1988 Healing: therapeutic touch. In Rankin - Box D (ed) Complementary therapies. Chapman Hall, London
Weiss S J 1986 Psychological effects of caregiver touch on incidence of cardiac dysrhythmia. Heart and Lung 15(5): 496–503
White J, Krippner S 1977 Future Science. Anchor Books, New York
Wolf S 1950 Effects of suggestion and conditioning on the action of chemical agents in human subjects – the pharmacology of placebos. Journal of clinical investigation 29: 100–109
Wright S M 1987 The use of Therapeutic Touch in the management of pain. Nursing Clinics of North America 22(3): 705–714
Wright S M 1991 Validity of the human energy field assessment form. Western Journal of Nursing Research 13(5): 635–648
Zefron L 1975 The history of the laying on of hands. Nursing Forum 15: 350–363

SELECTED CASE STUDIES ON THERAPEUTIC TOUCH

Bulbrook M J T 1984 Bulbrook's model of Therapeutic Touch: one form of health and healing in the future. The Canadian Nurse 80(11): 30–34
Heidt P R 1991 Therapeutic Touch – the caring environment. Journal of Holistic Nursing 9(3): 19–25
Krieger D 1979 The Therapeutic Touch. Prentice-Hall, New York
Macrae J 1979 Therapeutic Touch in practice. American Journal of Nursing 79(4): 663–665

3

Providing a conceptual framework for practice

Francis C Biley

Rogers' Science of Unitary Human
 Beings
Practical applications
Therapeutic Touch and Rogers' Science

■ *In the deeper reality beyond space and time, we may all be members of one body.*

SIR JAMES JEANS

Perhaps more than anybody else, Theresa Meehan has explored the concept of Therapeutic Touch (TT) from a Rogerian perspective. In other words, she has redefined the theoretical basis for an understanding of the technique. She (Meehan 1990a) notes how

> ... *the human being is a unitary phenomenon, energy fields are the fundamental units of the individual and environment, and both are in a process of continuous, simultaneous interaction and change.*

In order to understand that perspective, Martha Rogers' conceptual framework, the Science of Unitary Human Beings, needs to be explored.

Rogers' Science of Unitary Human Beings

Nursing in the United Kingdom has, since the 1970s, benefited from the influence of innovations that have arisen from North America. The nursing process, quality assurance, primary nursing, nursing research and the implementation of nursing models in practice are just a few of the topics that have gained wide application in this country. But there is one major area of study and practice that has been developed and is receiving considerable attention in North America which has yet to achieve anything more than a brief acknowledgement in the United Kingdom. The innovative work of Martha Rogers (1970, 1980, 1983, 1986, 1990) and other eminent nurses who have based their practice, education and research on Rogers' Science of Unitary Human Beings has remained in a state of relative obscurity in this country. New York University must have been an exciting place to be in the 1960s and 1970s. Dolores Krieger, a Professor in the Division of Nursing, and one time student of Martha Rogers (Malinski 1993) was beginning to understand and develop TT.

At the same time but as far as can be ascertained from the literature, independently from the work of Dolores Krieger, Martha Rogers, another Professor in the same Division, was beginning to develop what has come to be known as the conceptual framework of the Science of Unitary Human Beings. At the time and even 20 years on, this provided a radical vision of nursing reality. It was unique in providing a framework for nursing practice, education and research that promised a move away from the previously predominant medical model approach to the delivery of nursing care.

The framework provided a radical alternative to the traditional view of nursing which could be described as Cartesian, that is reductionistic, mechanistic and analytic, consisting of 'breaking up thoughts and problems into pieces and arranging these in their logical order' (Capra 1991). According to some, it has 'guided nursing out of a concrete, static, closed system world view' (Smith 1989) and, as a result, has challenged many preconceived ideas about nursing and beyond. Indeed, when the theoretical framework was first published, it was 'in clear contradiction to all the nursing theories in use at that time' (Sarter 1988a). Rather than just describing nursing as it is, the Science of Unitary Human Beings pushed forward and will continue to push forward the boundaries of nursing, suggested multiple therapeutic possibilities and provided a conceptual framework on which to base TT, amongst many other things and (perhaps most forward thinking and controversially) began to explore the possible implications of nursing in space (Christensen et al 1993). The Science of Unitary Human Beings and Rogerian nursing science offer a view of nursing that is in keeping with current developments in the understanding of the world, such as those suggested by Bohm (1980) and Pribram (1976).

Although the Science of Unitary Human beings is hardly recognised in this country, the scale of the influence of the framework on North American nursing is considerable. Every other year international conferences drawing delegates and presentations from all over the world are held in order to disseminate latest information. The New York based Society of Rogerian Scholars, which has its members in, for example, North America, Canada, Brazil, Germany and the UK exists in order to provide a forum for debate and to enable the exchange of ideas and views through local meetings and the quarterly newsletter the 'Rogerian Nursing Science News'. Significant textbooks dedicated to Rogers' work (notably Malinski 1986, Sarter 1988a, Barrett 1990a, Lutjens 1991) have been published. Rogers herself was a prolific author, and there is a wealth of other published material, with articles frequently appearing in significant nursing journals such as *Nursing Science Quarterly* and *Holistic Nursing Practice*. There is an annual journal entitled *Visions*, containing only work that explores issues raised by the Science of Unitary Human Beings.

Such is the significance of the Rogerian conceptual system that many notable nurse theorists, including for example Rosemary Rizzo Parse with her theory of 'Human Becoming' and Margaret Newman with her theory of 'Health as Expanding Consciousness', have based their work on the Science of Unitary Human Beings.

Martha Rogers was born in 1914, and between 1931 and 1954 received a wide nursing and academic education. After working in community health nursing she moved into higher education, eventually spending 21 years as Professor and Head of the Division of Nurse Education at New York University. In 1975 she retired and became Professor Emeritus in the same establishment (Falco & Lobo 1985). Considered to be a nurse leader and theorist whose significant contribution to the development of nursing theory, practice, education and research cannot be surpassed, she died on March 13 1994 at the age of 79.

It is perhaps a measure of the success of the Science of Unitary Human Beings that the curriculum on which undergraduate and postgraduate education is based is still, more than 20 years after her retirement, firmly grounded in her ideas. For those who may be interested, a more detailed exploration of her life history can be found elsewhere (Hektor 1989).

The seeds of the conceptual framework, the Science of Unitary Human Beings, were first seen in *Reveille in Nursing*, Rogers' second book, which was published in 1964. Six years later in 1970, Rogers published an expansion of this earlier work in *An Introduction to the Theoretical Basis of Nursing*. As already stated, there appear to be no explicit, direct links between the developmental work of Dolores Krieger and the formulation of the Science of Unitary Human Beings. However, as will be shown, there are surprising similarities in the proposals suggested by the two nurse academics, and the Science of Unitary Human Beings is now often used as the conceptual framework to support TT (Malinski 1993).

In the 20 years or so that followed the launch of the Science of Unitary Human Beings, considerable changes and developments were made to refine the conceptual framework (Rogers 1980, 1983, 1986, 1990). The original substantive work *An Introduction to the Theoretical Basis of Nursing* is very interesting to read and some would argue that it is a seminal text of historical significance (Lutjens 1991) which in years to come will assume the same importance and significance as Florence Nightingale's *Notes on Nursing*. While it would be an interesting exercise to chart these changes, it is beyond the scope of this chapter, which will give the reader only the most recent currently available definitions of the concepts subsumed under the Science of Unitary Human Beings. At the same time the immense value of these concepts will be emphasised by giving examples of their application in nursing practice and particularly TT, education and research.

The Science of Unitary Human Beings draws on a vast array of subjects that form the theoretical underpinning for the conceptual framework. These include, amongst others, anthropology, astronomy, mathematics (Daily et al 1989) and Einsteinian (post-Newtonian) physics (Sarter 1988a) and philosophy, including, amongst others, the work of Polanyi and de Chardin (Sarter 1988b)

It is these origins of the Science of Unitary Human Beings that differentiate the work of Martha Rogers from the work of most other nurse theorists and also perhaps Dolores Krieger, who developed ideas based on the insights developed from Eastern philosophical thought (Quinn & Strelkauskas 1993).

Such disciplines often use terminology in a very specific way and demand an in-depth knowledge and understanding of concepts that are very difficult to express and may be very alien to many British, and indeed North American, nurses. This could be one reason that the Science of Unitary Human Beings has received little emphasis in British nursing. It is certainly the reason why it has been called an 'outrageous nursing theory' (Thompson 1990), the complexity of which is 'difficult to understand' (Daily et al 1989) without some knowledge of, for example, quantum physics. However, if an attempt is made to try to understand the conceptual framework (and in some instances this might require prior background reading in some of these supporting subjects), readers will begin to realise why Rogers has been hailed 'a brilliant nurse theorist' and 'one of the most original thinkers in nursing' (Daily et al 1989), without whom it is 'difficult to imagine what nursing would look like today' (Barrett 1990b).

FIVE ASSUMPTIONS

In 1970, Rogers formulated five basic assumptions that describe human beings and their life process (Rogers 1970). These assumptions or 'building blocks', that have more recently been called 'postulates' (Malinski 1993) underpin the conceptual framework and consist of the concepts of:

Wholeness – in which the person is regarded as a unified whole possessing his/her own integrity with characteristics that are more than and different from the sum of the parts. In other words, an individual cannot be divided into, say, a series of systems or 'activities of living'. Such an approach is reductionist and fails to address the complexity that makes up the whole person.

Openness – where the individual and the environment are continuously exchanging matter and energy with each other. Rather than being of solid form, all things are made of energy and only appear solid.

Unidirectionality – where the life process exists along an irreversible space-time continuum. This is a concept that was abandoned in later writings.

Pattern and Organisation – which identifies individuals and reflects their innovative wholeness. For example, each person consists of energy that has a particular pattern and organisation and it is this that gives everyone their unique identity.

Sentience and Thought – which states that of all life, human beings are the only ones capable of abstraction and imagery, language and thought, sensation and emotion.

FOUR ELEMENTS

In a series of changes that illustrate the ever-developing nature of the Science of Unitary Human Beings over the years since 1970, four 'critical elements' emerged from these basic assumptions that describe the person and the life process (Cowling 1986). These critical elements are now regarded as 'basic to the proposed system' (Rogers 1986) and are

- energy fields

- open systems

- pattern and

- pandimensionality (Rogers 1991)

These four critical elements now form the basis of most of the detailed scientific and philosophical explorations of the Science of Unitary Human Beings and have replaced the previous five building blocks outlined above.

Energy fields are the 'fundamental unit of the living and the non-living' (Rogers 1986). They consist of the human energy field and the environment energy field. The human field is 'an irreducible, indivisible, pandimensional energy field identified by pattern and manifesting characteristics that are specific to the whole and which cannot be predicted from knowledge of the parts' (Rogers 1991). The environmental field is integral with the human field. Each environmental field is specific to its given human field. In other words, 'what has previously been regarded as solid matter now needs to be considered energy with ... the illusion of solidity' (Biley 1993). Although the human field is different from the environment field, they are integral and inseparable, and there can therefore be no exchange of energy between the different fields. It is this aspect of the Science of Unitary Human Beings that does not sit happily with the construction described by Dolores Krieger, who believes that there is an exchange of energy between a field with a healthy abundance of energy and a field of reduced energy that can be utilised when healing (Krieger 1979).

Open systems or openness describes the open nature of the fields, which allows for a continuous interchange of energy between the fields.

Pattern is the 'distinguishing characteristic of the energy field perceived as a single wave' (Rogers 1986), which gives identity to the field. Human behaviour can be regarded as manifestations of changing pattern (Alligood 1989).

Pandimensionality describes 'a non-linear domain without spacial or temporal attributes' (Rogers 1991). It is fundamentally different from the three-dimensional world which we usually view. Pandimensionality, (previously known as multidimensionality and prior to that in the early publications, four-dimensionality) is a term that would be used by Rogerian scientists to describe what might otherwise be labelled supernatural experiences or encounters, astral projection, spiritual events, déjà vu and so on.

THREE PRINCIPLES

Such a brief summary of the fundamental basis of the Science of Unitary Human Beings does not do justice to the concepts outlined and does little to explain them. An outline is necessary however, in order to place in context the key to the conceptual framework, the *Principles of Homeodynamics*. These principles 'postulate a way of perceiving unitary man' (Rogers 1970) and arise from these previous statements, giving 'fundamental guides to the practice of nursing' (Rogers 1990).

The *Principles of Homeodynamics* consist of the three principles of

- integrality
- helicy and
- resonancy

It is these aspects of the Science of Unitary Human Beings that have had more direct relevance for nursing practice, research and education than the descriptions of human beings and the life process that have previously been outlined.

Integrality describes the 'continuous mutual human field and environmental field process' (Rogers 1990), suggesting that human and environment energy fields are open systems that continuously pass through one another (Alligood 1989).

Helicy describes the 'continuous innovative, unpredictable, increasing diversity of human and environmental field patterns' (Rogers 1990), the 'continuous creative development and evolution of the human-environmental fields' (Gueldner 1989).

Resonancy describes the 'continuous change from lower to higher frequency wave patterns in human and environmental fields' (Rogers 1990).

In order to gain a greater understanding of these principles rather than try to explain the precise way in which these words have been used (a process that is extremely difficult if not impossible), it is probably more useful to see how different authors have interpreted them in their research and practice.

Integrality

The Principle of Integrality, where the human field is integral or at one with its environmental field (Schodt 1989), was studied by McDonald (1986), who stated that if there is a continuous mutual human field and environmental process, changes in one field will bring about changes in the other. In other words 'researchers should be able to demonstrate a relationship between a nurse-initiated modification in a person's environment and an alteration in that person's state of being' (McDonald 1986). In order to examine whether such nursing concepts can be brought down the ladder of abstraction to an operational level (Smith 1986), McDonald tested whether an alteration in the colour of light (the environmental field) could bring about a reduction in rheumatoid arthritis pain in the left hands of 60 female volunteers (human fields) and found that blue lightwaves were related to a reduction in the experience of pain. Such findings re-emphasise a whole new world of nursing practice possibilities. It has been suggested that other interventions, such as music

(Smith 1986), humour or meditation can be used by the nurse to promote positive human-environment field patterning (Malinski 1986).

Helicy

The Principle of Helicy was explored by Floyd (1983) who made the prediction that the amount of wakefulness and the number of sleep–wakefulness cycles increase (in other words, there will be an increasing diversity of human and environmental field patterns) when 'persons experience a deviation in the rhythmic relationship with their environment' (Floyd 1983). Shift rotation and admission to hospital were two variables that were identified as possibly causing diversity in field patterning. 60 shift workers and 35 hospital in-patients took part in the study which found that those with diversity in field patterning did indeed sleep less (0.3 hour and 1.0 hour less per cycle for shift workers and hospital inpatients respectively) than those without such diversity. The number of sleep–wakefulness cycles did not increase for hospital inpatients although it did increase for shift workers. Floyd (1983) indicated that this kind of information may be useful when planning 'optimum systems for around-the-clock delivery of nursing ...'. It also indicated that diversity in human and environmental field patterns occurs for individuals on admission to hospital and that nurses should take steps to promote positive human-environmental field re-patterning. In another study that examined the principle of helicy, Rawnsley (1986) found that perceptions of the speed of time was slower for those who were dying and that this could be an explanation for the lack of patience often experienced by the terminally ill.

Resonancy

The Principle of Resonancy, in which it is suggested that there is a continuous change from lower to higher frequency wave patterns in human and environmental field patterns, was studied, for example, by Butcher & Parker (1988). They postulated that the technique of guided imagery could promote resonancy, and that the change from lower to higher frequency patterning could be a theoretical explanation for feelings such as relaxation and timelessness. 60 adult subjects were assigned to an experimental group (who listened to a guided imagery tape) or a control group (who listened to an educational tape). The experimental group had a significantly lower score on the specially developed Time Metaphor Test than the control group, indicating a greater sense of timelessness or higher frequency wave patterning. Subjective opinions from both groups confirmed these results. The authors state that these results illustrate 'the potential of the Science of Unitary Human Beings for providing a scientific rationale of the use of pleasant guided imagery in nursing practice' (Butcher & Parker 1988) which can promote harmony, relaxation and wellbeing.

It has been stated that Rogers' Science of Unitary Human Beings offers a view of nursing that is consistent with the currently prevailing world of holism and ecological concern (Cowling 1986) and in a sense was, perhaps still is, ahead of its time. If nursing is to firmly embrace these ideologies it needs to firmly embrace the Science of Unitary Human Beings with all its implications for nursing practice, education and research. The above examples of empirical explorations of the principles of homeodynamics show how the Science of Unitary Human Beings can provide the philosophical framework for the development of research studies that influence the development of nursing practice. But there are also other areas where the Science of Unitary Human Beings has had a direct application to nursing practice and education.

Practical applications

Many examples have been given of the direct application of the conceptual framework in nursing practice. Bradley (1987) and Hover-Kramer (1990) promote the importance of the concept of energy fields and its potential in techniques such as TT. Whelton (1979) presented a comprehensive assessment and care plan based on the Science of Unitary Human Beings. It was shown to be useful in guiding nursing intervention and predicting outcomes in the examples given, that is, the care of a patient with decreased cardiac output, diabetes and hypertension and in the care of another with a recurrent meningioma.

Another assessment tool to be used in Rogerian nursing practice has been developed by Barrett (1988) who states that nurses need to assess 'pattern manifestation' and to promote 'deliberative mutual patterning' (Barrett 1990c). The nursing care of an adolescent with a 'borderline personality disorder' has been described by Thompson (1990) who used the conceptual framework to describe the inter-personal processes of transference and counter-transference that existed.

Further explorations of patient care scenarios using the Science of Unitary Human Beings are given by Meehan (1990b) who describes caring for a man with pain due to metastatic cancer and Madrid (1990) who gives a moving account of successful deliberative mutual patterning in the care of a patient who was in considerable discomfort due to pain, hospitalisation and gastrointestinal bleeding.

Therapeutic Touch and Rogers' Science

In recent years, the Science of Unitary Human Beings has proved to be a relevant framework for the exploration and explanation of TT. Several TT practitioners and theorists, such as Quinn (1984) and Meehan (1990a), have used the Science of Unitary Human Beings as a means of providing a rationale and an explanation for the practice of TT and the foundation for most of the TT studies (Quinn & Strelkauskas 1993). Although this work is regarded as very influential, it has not been without its difficulties and is only in the very early stages of development and understanding.

As already stated, Krieger developed TT on a basis taken from Eastern philosophies and systems of medicine. Energy, or alternatively prana or ch'i, is thought by Krieger to flow from the healthy TT practitioner who has an abundance, to the ill person with depleted levels of energy. Meehan (1990a) suggested that this concept is incompatible with the Science of Unitary Human Beings, where there is never any 'to-from exchange or interaction between practitioner and client' (Malinski 1993).

Quinn has therefore re-conceptualised TT and, rather than using a pure Krieger-Kunz method and rationale, has begun to explore what has become known as the Rogerian-Quinn framework for TT (Olson et al 1992).

Meehan (1990a) re-defined TT 'as a knowledgeable and purposive patterning of patient-environment energy field process in which the nurse assumes a meditative awareness and uses her/his hands as a focus for the patterning of the mutual patient-environment energy field process', suggesting that the Krieger term 'energy exchange' should be changed to the Rogerian term 'energy process' and that 'energy transfer' should become 'mutual process'. She also stated that TT should not be seen as derived from 'laying on of hands'.

Other differences between TT as perceived from a Krieger-Kunz perspective and as perceived from a Science of Unitary Human Beings perspective have been identified (Malinski 1993). Some of these differences are, for example, that from a Rogerian perspective, TT is not seen as a nurse-patient interaction because the nurse should be regarded as being part of the patient's environment. It is therefore a nurse-environment process. Additionally, Rogers prefers to employ words such as participation rather than motivation or intentionality because the 'Therapeutic Touch practitioner is neither an instrument of nor a conduit for a higher healing power' (Malinski 1993).

This exploration of the Science of Unitary Human Beings and the re-conceptualisation of TT is by no means fully comprehensive (more extensive reviews, primarily of the Science of Unitary Human Beings, can be found in Barrett 1990a and Daily et al 1989). Nevertheless, the aforementioned literature provides evidence that points to the expanding adoption of the Science of Unitary Human Beings as a framework to guide nursing practice and the practice of TT in both the relative present and the relative future. Even so, a considerable amount of work needs doing before a complete re-conceptualisation of TT is achieved.

As this chapter has shown, trying to describe the Science of Unitary Human Beings without moving too far away from (and perhaps therefore losing the essential meaning of) the terminology employed by Martha Rogers and other Rogerian scientists is a very difficult task. An understanding of the Science of Unitary Human Beings requires a broadly based unitary knowledge-base and education.

Rogers stated that 'education is for the future, yesterday's methods will not suffice for tomorrow's needs' (Rogers 1961). If nursing education is to respond positively to the need to educate nurses for the future, to adapt successfully to the changing needs and demands of a New Age, ecologically-concerned society, then it will need to widen its knowledge-base and adopt a framework such as that described here.

PRACTICAL NOTE

For those interested, examples of Rogerian-based care plans can be found in Biley (1993), Mills & Biley (in press) and Madrid (1990). Further information on Rogerian Nursing Science and the Society of Rogerian Scholars can be obtained from:

Francis C Biley
School of Nursing Studies
University of Wales College of Medicine
Heath Park
Cardiff
Wales
United Kingdom
Telephone: 0222 743734

REFERENCES

Alligood M R 1989 Rogers' theory and nursing administration: a perspective on health and environment. In: Henry B, Arndt C, DiVincenti M et al (Ed) Dimensions of nursing administration: theory, research, education and practice. Blackwell Scientific, Boston

Barrett E A M 1988 Using Rogers' science of unitary human beings in nursing practice. Nursing Science Quarterly 1: 50–51

Barrett E A M 1990a Visions of Rogers' science-based nursing. National League for Nursing, New York

Barrett E A M 1990b Preface. In: Barrett E A M (Ed) Visions of Rogers' science-based nursing. National League for Nursing, New York

Barrett E A M 1990c Rogers' science-based nursing practice. In: Barrett E A M (Ed) Visions of Rogers' science-based nursing. National League for Nursing, New York

Biley F C 1993 Energy fields nursing: a brief encounter of a unitary kind. International Journal of Nursing Studies 30(6): 519–525

Bohm D 1980 Wholeness and the implicate order. Routledge, London

Bradley D B 1987 Energy fields: implications for nurses. Journal of Holistic
    Nursing 5(11): 32–35
Butcher H K, Parker N I 1988 Guided imagery within Rogers' science of
    unitary human beings: an experimental study. Nursing Science Quarterly
    1: 103–110
Capra F 1991 The Tao of Physics. Shambhala, Boston
Christensen P, Sowell R, Gueldner S H 1993 Nursing in space: theoretical
    foundations and potential applications within Rogerian Science. Visions
    1(1): 36–44
Cowling W R 1986 The science of unitary human beings: theoretical issues,
    methodological challenges, and research realities. In: Malinski V M
    (Ed)Explorations on Martha Rogers' science of unitary human beings.
    Appleton-Century-Croft, Conneticut
Daily J S, Maupin J S, Slattery M C, Schmell D L and Wallace T L 1989
    Martha E Rogers: unitary human beings. In: Marriner-Tomey A (Ed)
    Nursing theorists and their work (2nd ed). Mosby, New York
Falco S M and Lobo M L 1985 Martha E Rogers. In: George J B (Ed) Nursing
    theories: the base for professional nursing practice (2nd ed) Prentice-Hall,
    New Jersey
Floyd J A 1983 Research using Rogers' conceptual system: development of a
    testable theorem. Advances in Nursing Science 5(2): 37–48
Gueldner S H 1989 Applying Rogers' model to nursing administration:
    emphasis on client and nursing. In: Henry B, Arndt C, DiVincenti M,
    Marriner-Tomey A (Eds) Dimensions of nursing administration: theory,
    research, education and practice. Blackwell Scientific, Boston
Hektor L M 1989 Martha E Rogers: A life history. Nursing Science Quarterly
    2(2): 63–73
Hover-Kramer D 1990 Energy fields: implications for the science of human
    caring. Imprint 37(3): 81–2
Krieger D 1979 The therapeutic touch: how to use your hands to help or
    heal. Prentice-Hall, New Jersey
Lutjens L R J 1991 Martha E Rogers: the science of unitary human beings.
    Sage, London
McDonald S F 1986 The relationship between visible lightwaves and the
    experience of pain. In: Malinski V M (Ed) Explorations on Martha E Rogers'
    science of unitary human beings. Appleton-Century-Crofts, Connecticut
Madrid M 1990 The participating process of human field patterning in an
    acute-care environment. In: Barrett E A M (ed) Visions of Rogers' science-
    based nursing. National League for Nursing, New York
Malinski V M 1986 Explorations on Martha E Rogers' science of unitary
    human beings. Appleton-Century-Crofts, Connecticut
Malinski V M 1993 Therapeutic touch: the view from Rogerian nursing
    science. Visions 1(1): 45–54

Meehan T C 1990a Theory development. In: Barrett E A M (ed) Visions of
Rogers' science-based nursing. National League for Nursing, New York

Meehan T C 1990b The science of unitary human beings and theory based
practice: Therapeutic Touch. In: Barrett E A M (ed) Visions of Rogers'
science-based nursing. National League for Nursing, New York

Mills A, Biley F C (in press) Adam: a case study in Rogerian nursing
practice. Nursing Standard.

Olson M, Sneed N, Bonadonna R, Ratcliff J, Dias 1992 Therapeutic Touch
and post-Hurricane Hugo stress. Journal of Holistic Nursing 10: 120–136

Pribram K H 1976 Problems concerning the structure of consciousness. In:
Globus G (Ed) Consciousness and the brain. Plenum, New York

Quinn J F 1984 Therapeutic Touch as energy exchange: testing the theory.
Advances in Nursing Science 6(2): 42–49

Quinn J F, Strelkauskas A J (1983) Psychoimmunologic effects of Therapeutic
Touch on practitioners and recently bereaved recipients: a pilot study.
Advances in Nursing Science 15:(4) 13–26

Rawnsley M M 1986 The relationship between the perception of the speed of
time and the process of dying. In: Malinski V M (ed) Explorations on
Martha Rogers' science of unitary human beings. Appleton-Century-
Crofts, Connecticut

Rogers M E 1961 Educational revolution in Nursing. Macmillan, New York

Rogers M E 1964 Reveille in Nursing. Davis, Philadelphia

Rogers M E 1970 An introduction to the theoretical basis of nursing.
F A Davis, Philadelphia

Rogers M E 1980 Nursing: A Science of Unitary Man. In: Riehl J P and Roy C
(Eds) Conceptual Models for Nursing Practice, Second Edition. Appleton-
Century-Crofts, New York.

Rogers M E 1983 Science of unitary human beings: a paradigm for nursing.
In: Clements I W, Roberts F B (Eds) Family health: a theoretical approach
to nursing care. Wiley, New York

Rogers M E 1986 Science of unitary human beings. In: Malinski V M (Ed)
Explorations on Martha Rogers' science of unitary human beings.
National League for Nursing, New York

Rogers M E 1990 Nursing: science of unitary, irreducible, human beings:
update 1990. In: Barrett E A M (Ed) Visions of Rogers' science-based
nursing. National League for Nursing, New York

Rogers M E 1991 Glossary. Rogerian Nursing Science News 4(2): 6–7

Sarter B 1988a The stream of becoming: a study of Martha Rogers' Theory.
National League for Nursing, New York

Sarter B 1988b Philosophical sources of nursing theory. Nursing Science
Quarterly 1: 32–59

Schodt C M 1989 Parental-fetal attachment and couvade: a study of patterns
of human-environment integrality. Nursing Science Quarterly 2(2): 88–97

Smith M J 1986 Human-environment process: a test of Rogers' principle of integrality. Advances in Nursing Science 9(1): 21–28

Smith M J 1989 Four dimensionality: where to go with it. Nursing Science Quarterly 2(2): 56

Thompson J E 1990 Finding the borderline's border: can Martha Rogers help? Perspectives in Psychiatric Care 26(4): 7–10

Whelton B J 1979 An operationalisation of Martha Rogers' theory throughout the nursing process. International Journal of Nursing Studies 16: 7–20

4

The essentials of practice

Jean Sayre-Adams   Steve Wright

Explorations
Several steps
Related issues
Factors affecting success

■ *'I can't believe that' said Alice.*
'Can't you?' the Queen wailed in a pitying tone.
'Try again; draw a long breath, and shut
your eyes.'

LEWIS CARROLL *Through the Looking Glass*

This chapter will explore the practical application of TT as well as ideas about learning to do it. It is important, however, that anyone reading this book recognises from the outset that the practice of TT to cannot be learned from books. It is not possible to read about principles and then assume that they can immediately be put into practice. We strongly recommend anyone wishing to practice TT to pursue a recognised course of study, some suggestions for which will be given in the last chapter. The purpose of this section is to provide an explanation of the essential principles of the practice of TT – an understanding of which is fundamental to effective practice.

Explorations

EXERCISE 1 **THERAPEUTIC TOUCH SELF-KNOWLEDGE TEST**

1. Bring your attention to your hands. You may perhaps explore one hand with the other. Stretch and wiggle the fingers and wrists. Perhaps you may want to rub your palms together. Become more aware of your hands and how they serve you.

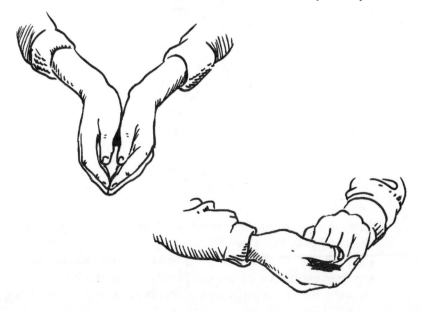

2. Then sit comfortably with both feet on the ground and simply place your hands so that the palms face each other. Hold your elbows away from the trunk of your body and do not rest your lower arms in your lap. Now bring your palms as close together as you can without having them touch each other, so that they are perhaps one-eighth to one-quarter inch apart.

3. Separate the palms of your hands by about two inches and then slowly bring them back to their original position.

4. Now separate your palms by about four inches and, again slowly bring them back to their original position.

5. Repeat this procedure. However, this time, separate your palms by about six inches. Keep your motions slow and steady. As you return your hands to their original position, notice if you begin to feel a build up of pressure between your hands or if you feel any other significant sensation

6. Once again separate your palms, this time until they are about eight inches apart. Do not immediately return your hands to their original position. Instead, as you bring you hands close together, experience the pressure field you have built up by stopping for a moment every two inches or so, and slowly

trying to compress the field between your hands. You may experience this as a 'bouncy' feeling.

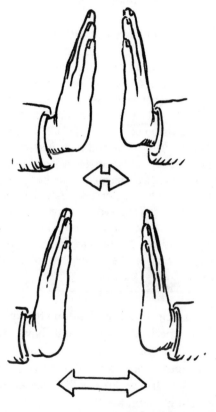

7. Spend the next full minute in experiencing this field between your hands and try to determine what other characteristics of the field you feel besides the pressure and the bounciness or elasticity.

The sensations that people will 'feel' are almost as varied as people themselves. Practice will help you become clear on how to express yourself and understand these cues. These sensations and cues will be very subtle but distinct.

(revised from Krieger 1979)

No one to date has been able to measure the exchange of energy. However, people who have had the experience of this energy exchange (either by feeling or sometimes seeing energy fields) have no doubt of its existence.

The following incident leaves us with even more questions:

> I had finished the first day of a two-day introductory TT workshop. We had finished the day with an experience and discussion about energy fields in which we had been discussing and experiencing the colours yellow, orange and red. Leaving that all behind, I then went with a friend to pick her children up from school and we moved into talking with the children about their day and continued after we got home focusing on the children's issues, not thinking or talking about the TT class at all. One of the girls disappeared for half an hour and when she returned she brought with her a picture which she showed to me. I asked her who it was, and she told me it was me. I then commented on how pretty I appeared to her and then asked her what the lines around me were. She told me that it was my 'pattern' and that my pattern looked pretty to her when it was that colour (see back cover)- yellow, orange and red.

There are many exercises that can be done to experience these fields, and the kinds of experiences people will have are as varied as the people. However, some of the common sensations are heat, warmth, coolness, 'like a breeze', tingling, bounciness, springiness, pressure, magnetic pull, sticky, stringy, 'like pulling toffee', soft chewing gum', 'steel bars', 'crackling', 'vibrations'.

EXERCISE 2A USING YOUR FIELD AS A DATA BASE

Now that you have had one experience of how your energy field extends beyond your skin, here is another exercise that may be done to elicit more information from your field and the environment around you.

This exercise works best when done in groups of people.

1. Have the members of the group sit in a circle with a foot or two between each other.

2. Lightly close your eyes and, in your mind's eye, begin to get a sense of the energy field surrounding your entire body.

3. Now, in your mind's eye, focus your attention on the area around your right shoulder and get an idea of what that space feels like. THEN

4. Using your intention, 'extend' that space beyond your right shoulder so that it reaches towards the person on your right hand side.

5. It may be, at this time, that various impressions may arise in your mind – these may be perceived as colours, symbols, images, scenes, emotions, smells, tastes, etc., that may or may not have any meaning for you. Just notice them without trying to understand.

6. Tuck the information you received, both from yourself as you were getting a sense of the field surrounding your entire body and from the person on your right, away in your mind or jot it down on a piece of paper so you can remember it. Keep the experiences from the first and second part of this exercise separate.

———————————————

Now, go onto the next exercise.

EXERCISE 2B INTENSIFYING THE FIELD EFFECTS

Members of the group should stay in their circle for this exercise.

For this exercise each person will need a piece of pure cotton wool.

1. Lay the piece of cotton wool in one hand put the other hand over it without touching.

2. In your mind's eye, reach down with your upper hand into your piece of cotton wool, through to your hand underneath.

3. Rub your piece of cotton wool over your hands, arms and your face – smell it, experience it fully. THEN

4. Take your piece of cotton wool and pass it to the person on your left.

5. Holding the piece of cotton wool received from the other person in one hand, put your other hand over it, but not yet touching, and with your mind's eye 'reach' down into it – then begin to touch it and feel it – all the while being aware of any impressions that arise in your mind. Again, be aware of colours, symbols, images, scenes, emotions, smells, tastes, etc. Do not question the impressions, no matter how illogical, odd, or unfamiliar they may be. Just accept them and tuck them away in your mind or jot them down on a piece of paper. Keep the experiences between the first and second parts of this exercise separate.

The first parts of the two exercises above may give you information about yourself as you intensified your field; the information from the second part of the exercises may have been elicited from the field of the person on your right. Share among yourselves and in the group the likenesses that may have emerged.

(revised from Krieger 1979)

The kinds of experiences people will have vary a great deal. Below are a small sample of ones noted by teachers from different TT classes in several countries.

1st participant   'I felt like I was flying through the air. I saw a Baltimore Oriole (North American bird) and heard flute music. I was bicycling down a green lane'.

2nd participant (whom the 1st had not known or talked to before) 'I just returned from Egypt where I listened to someone playing the flute near the pyramids. I am flying to Baltimore tomorrow and my girlfriend has told me she has discovered a lovely, new, green country lane for us to bicycle on'.

Again in another workshop:

1st participant   'I was overcome with a feeling of compassion'.

2nd participant (with tears in her eyes) 'I felt as if I was being given a huge, all-encompassing hug'.

3rd participant   'I smelled the smell of an oak fire' (an oak fire had memories of love for her).

Again:

1st participant   'I felt heavy, dark, nothing'.

2nd participant   'I was determined that the person was not going to get any information from me'.

By doing these exercises you might have found out that:

- You don't stop at your skin: there is a field beyond your skin which can be experienced.

- You can use this field as a database: when you turn your attention to this personal space, you find that it can help you to elicit additional cues about your environment.

- You can intensify these fields so that they are more perceptible to you.

Several steps

Almost every article on TT lists 4, 5 or 6 steps on how to do TT. It needs to be emphasised that while most teachers do use a step by step method when actually teaching TT, these steps are not sequential nor necessarily done in a linear fashion. They flow into one another and it is often impossible for someone watching to distinguish one step from another.

These steps or phases are as follows:

CENTRING

Centring is achieved by shifting awareness from an external to an internal focus, becoming relaxed and calm and making a mental intention to assist the patient.

Practitioners describe centring as 'a method for:

1. disciplining attention

2. achieving calm, and

3. establishing receptivity'. (Lionberger 1985)

It is focusing on the here and now, placing oneself in a calm, alert, open state. It involves having a clear sense of oneself as a unitary whole and differs from concentration or paying attention in that it is not associated with mental effort. It brings the patient and practitioner into harmony with one another and so mobilises the TT process. Centring is paramount to proceeding with the remaining phases and is necessary for the intervention to be effective. Being unable to become or stay centred causes the practitioner to feel depleted of energy, tired and to be ineffectual (Sayre-Adams 1993). Centring is the repatterning of the practitioner's energy field in the direction of expanded consciousness.

> Remember that TT is an energy field interaction between you and another person. If you start a treatment with an energy field disturbed by being upset, anxious or in a negative frame of mind, the disturbances in your energy field tend to amplify the disturbances within the patient's field.

Students describe a centred place as being 'calm', 'peaceful', 'integrated', 'aligned', 'grounded', 'whole', 'at one' and 'in harmony' with their environment. The words used are almost as varied as the number of students. Equally as varied are the ways in which different practitioners arrive at their centre. Many will find that a meditation practice is helpful in learning to centre quickly and stay centred. (See the section on meditation later in this chapter.)

EXERCISE 3 A CENTRING TECHNIQUE

If you have not found your way yet, you may try the exercise below. This is one that has been used in TT classes for many years.

1. Sit comfortably and close your eyes.

2. Inhale and exhale deeply.

3. Focus your mind on some image in nature, such as a tree or a mountain, which brings you a sense of peace. If you become distracted, gently bring your mind back to your image of peace. Remember not to tighten up or try to force your mind to be still. Just maintain a calm but firm intent to keep focused on the image.

ASSESSMENT

Nurses and many other health care workers are taught to assess patients from the first contact. They look at X-rays, laboratory results, vital signs, skin tone and colour and so on. In some specialist units each system may be assessed, e.g. cardiac, respiratory etc. Assessment may also include how the patient walks, tone of voice, grooming or body language and so on. The assessment of the energy field is just one more piece of information about the patient's general condition. It is not a diagnostic tool. The imbalances found in the energy field by the assessment may or may not correspond to the symptoms of the patient.

The person carrying out the assessment needs to be relaxed, the mind open. The assessment or scanning is done by holding the hands three to five inches from the physical body of the patient and moving them throughout the complete field from head to toe to find the imbalances or differences in the field. A balanced field might have a soft, rhythmic, warm feel. Imbalances are highly subjective and will be perceived as subtly as the feeling in the hands in the TT self-knowledge test above. Some of the words used by students to describe these imbalances are 'warmth', 'cold', 'pins and needles', 'void', 'empty', 'heavy' and 'pressure'. More experienced practitioners also perceive information through intuitive and somatic clues rather than through the hands, though the hands are still the focal point for these clues. For example, the person practising TT may simply feel that they 'know' where the imbalance in the energy field is. Or they may feel it in a corresponding place in their own body – e.g. a patient's headache may be experienced as a pain in the head of the practitioner.

Most often the imbalances that are picked up in the assessment phase are one or a combination of the following:

- Temperature differentials, such as a sense of heat or cold

- Pressure, or feelings of congestion in the energy flow

- Changes in or lack of synchronisation in the intrinsic rhythmicity of the patient's energy field

- Localised weak electric shocks or tingly feelings

(Krieger 1993).

The clues are qualitative and difficult to describe precisely. Because assessing is always an ongoing process, it should not be a cause of concern if all the cues are not picked up the first time. Assessment and treatment of the patient may be carried out in any position he or she feels comfortable with. The following tips are given to ensure a successful assessment:

EXERCISE 4 SUGGESTIONS FOR THE ASSESSMENT

Below are step by step suggestions for the assessment. Either try these
or develop your own.

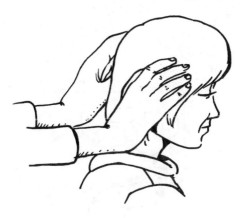

The following section is based on the work of Macrae (1987).

1. Ask the patient to sit on a stool or sideways on a chair so that
   his/her back is unobstructed.

2. Sit, kneel or stand to the side of the patient so that one of your
   hands – either the right or the left – extends in front of the
   patient and the other is behind the patient's back.

3. Pass your hands gently through the patient's energy field
   (about 3 to 5 inches from the surface of the skin), palms facing
   the body, starting from the top of the head and moving down
   to the feet. When the hand at the back reaches the patient's
   hips, keep it there while the other hand passes all the way
   down the legs to the feet.

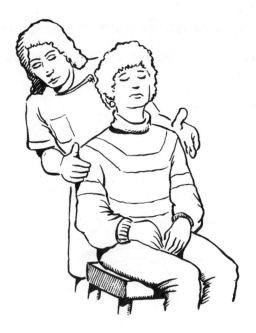

or

1. Ask the patient to sit on a stool or sideways on a chair.

2. Sit, kneel or stand in front of the patient; then pass your hands, parallel to each other with palms facing the patient, through the patient's energy field from head to foot.

3. Assess the back of the patient and pass your hands, in the same manner, from the patient's head down to the hips.

More tips:

1. Make all your movements down from top to bottom.

2. Your complete assessment should not exceed 60–90 seconds.

3. Do not linger on a spot while trying to decide what you feel. If in doubt, repeat the whole assessment again (starting with centring) from the beginning to ascertain the correctness of your cues.

4. If you lose your centre, stop, recentre yourself and start from the beginning.

5. Feel the arches of the feet to see if the energy coming from them is equal. If not, clear the blockage before you go any further.

6. Do not try too hard. This is an intuitive process and should be allowed to unfold.

CLEARING

In this step, the hands are used to facilitate a symmetrical and rhythmical flow of energy through the field. It is done by sweeping the hands just above the body, downward. The entire body is covered, with concentration over the areas where an imbalance or blockage was identified during the assessment.

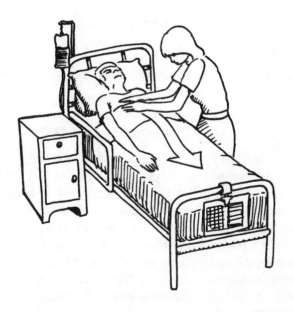

In Krieger's (1979) view, it is in this phase that the patients mobilise their own resources so that self healing can occur. It is commonly reported that responses indicative of deep relaxation occur most frequently during this phase. If the patient is experiencing anxiety, discomfort or pain of a physical or emotional nature, it is at this point that these symptoms may diminish. A visible or audible relaxation response, e.g. a deep breath or sigh, peripheral flush, deepening of the voice or other signs of physical relaxation will often be observed. Many practitioners believe this is not a separate step but part of the balancing or repatterning phase. Certainly the assessment, clearing and repatterning are performed concurrently and repetitively.

The act of clearing frees blocked energy. It can also be used to:

- facilitate energy flow
- break up energy patterns, such as congestion
- reestablish rhythm in the patient's energy field
- cool, in situations such as elevated temperature
- move patterned energy within the field
- knit together energy field.

It is intentionality, not hand movement, that is important (Krieger 1993).

TREATMENT – BALANCING OR REPATTERNING

Once a symmetrical flow of energy has been facilitated in the energy field through clearing, then the practitioner returns to the places where the imbalances were felt in the assessment and repatterns the field using intention and clear visualisation. For instance, if the imbalance was perceived as cold, the practitioner would attempt to warm it and vice versa. If a pressure is felt, an attempt would be made to reduce that pressure, if congestion, an attempt would be made to loosen and remove that congestion and so on. These changes or repatterning are accomplished through the appropriate direction, modulation or guidance of energy from one area of the energy field of the patient to another or from the environmental field to the patient.

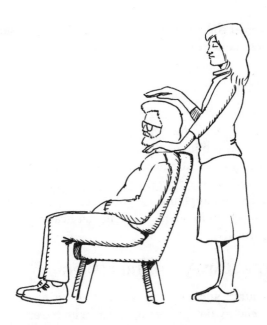

During this phase as in all others, the practitioner continues to sense their own field continually, simultaneously and mutually as a part of the patient's field, as well as the environmental field. Ultimately, there is really no distinction when fields are in contact.

| EXERCISE 5 | INTENTIONS THAT YOU MAY HAVE FOR THE REPATTERNING PHASE |

- the facilitation of energy flow

- the stimulation of energy flow

- the mobilisation of congestion or pressure in the energy field

- the dampening or quieting of energetic activity

- the synchronisation of rhythmicity in the energy flow.

Although these phases are usually taught sequentially to beginners, with experienced practitioners they become more dynamic and are often performed concurrently and repetitively. It is often difficult for an observer to notice distinct phases (Sayre-Adams 1994).

EVALUATION

A treatment can last anything from a few seconds to half an hour, and is 'finished' when the repatterning or rebalancing is complete. This is perceived intuitively by experienced practitioners or when a reassessment shows that balancing is achieved. A shorter time will always be needed when working with babies, frail or debilitated patients, and when applying TT to the head area. Pregnant women, patients in shock and patients with mental health problems are also often more sensitive.

EXERCISE 6 SUMMARY OF TREATMENT

1. Do your preliminary relaxation and centring exercise. Feel your energy field becoming at one with the energy field of your patient and the rest of the environment.

2. Place your intent upon the wholeness and wellbeing of the patient.

3. Gently massage the patient's neck and shoulders.

4. When you are centred and the patient is relaxed, assess the quality of their energy field.

5. Clear away any loose congestion (thickness, heat, pressure) making sure that the energy is flowing freely through the feet.

6. Loosen any obstructions (cold or blank areas) so that the field is open and flowing.

7. Fill in depleted areas, where you feel a pulling sensation.

8. Stop immediately and recentre yourself if you become drained of your own energy or if you absorb the patient's problems.

9. Reassess the energy field of the patient from time to time and clear away any congestion that may have surfaced.

10. Observe the patient for signs of relaxation (e.g. slower and deeper breathing, flushed face, muscle twitching).

11. If appropriate, ask for feedback from the patient and modify the treatment accordingly.

12. Conclude the treatment when the deficits have been filled in and the energy field is flowing in an open, even manner.

13. Ask the patient to sit or lie quietly for 20–30 minutes. These steps are not always performed sequentially. The more experienced the practitioner, the more these steps become one continuous, flowing movement.

(revised from Macrae 1987)

Macrae further suggests that a number of different techniques can be applied to different types of energy imbalance, congestion or deficit (see Table 4.1).

Table 4.1 *Suggested treatment for different types of energy imbalance (revised from Macrae 1987)*

Type of imbalance	Cue	Treat by	Be careful to
Loose congestion	Heat, thickness, heaviness, pressure	Clearing the energy field in a sweeping downward motion	Stay integrated so that you do not absorb the congestion, make sure the feet are not blocked
Energy deficit	Open hole drawing or pulling	Transferring energy into area until it feels full	Stay centred and relaxed. Reassess the field often – watch for signs of overload
Local imbalance	Pins and needles, static, discord	Smoothing the area until it feels in harmony with the rhythm of the entire field	Always visualise, or think of the energy field as being in balance
Tight congestion	Coldness, blankness	Physically touching the patient into the area to penetrate and loosen congestion, and then treat for loose congestion	Avoid force or strain – watch for signs of release both physical and emotional

Two practitioners working together

If there are two practitioners available, it is both pleasant and effective to work together on the same patient. In this case, both practitioners perform a complete assessment after they have centred and felt themselves to be in harmony with each other and the patient. Then the repatterning is carried out as if they were one unit, with the patient between them.

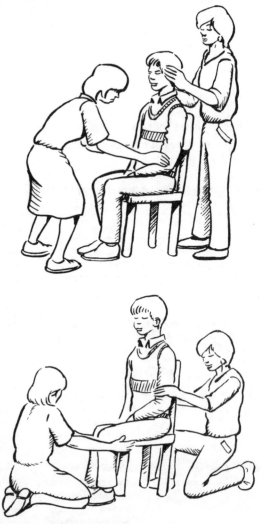

When I first met Mr J, a 61-year-old farmer, and invited him to come into the treatment room, I thought I would have to stop him falling over. His gait was unsteady but he did not require assistance.

Mr J told me that over the last 15 years his balance had progressively deteriorated to such an extent that, with very few exceptions, he fell every day. Some of the falls had resulted in a broken nose and arm, and cuts and bruises of varying severity. He stated that many people thought he was drunk and his walk certainly gave that impression. He stated his problems were due to a hereditary disease, that he was not receiving any medication and that his doctors were unable to do anything to help him.

Before commencing with TT, I centred myself by visualising an oak tree with its leaves shimmering in a gentle breeze and inwardly calming myself by taking a few deep breaths. I visualised a universal power of which I became a part. This personal method of centring is one which I find effective.

With Mr J lying on his back, I began TT. His initial assessment revealed an overall feeling of coolness in his energy field. As I had not felt any particular energy block or areas of deficit, my intuition led me to gently move towards Mr J's abdominal area and start the rebalancing. I imagined a warm soothing light encompassing Mr J. I again moved my hands in a head to toe direction to reassess and continued the rebalancing, moving my hands in a continuous, flowing manner. I continued this for approximately 10 minutes, until a feeling of warmth replaced the coolness. The same feeling of coolness was felt down the spinal column. This too was replaced with warmth after the rebalancing or repatterning.

Mr J had had no sensations but stated 'inwardly something is happening' and 'is it possible to feel better immediately?' His gait was unchanged.

The following week, he returned and was over the moon with how well he felt. He called me the lady with the oilcan – his joints were no longer stiff and he had not fallen all week. He was able to negotiate the steps in his house without having to hold onto the doorframe and he could put his shoes on without having to sit down.

On the second and third visits there were still areas of coolness in the energy fields around his legs and spinal column or static type shocks. A repatterning was done on the third visit and Mr J experienced a 'sensation travelling down my body'.

Three weeks later, he had his fourth and final TT treatment. At this time his energy field felt complete and he had not fallen for six weeks. His unsteady gait was still evident, but was not as exaggerated and he felt better able to cope. I believe TT improved Mr J's quality of life, enhanced his feeling of wellbeing and reduced his risk of physical injury through falls.

Related issues

MEDITATION

Meditation is often misunderstood. Perhaps this is because there are so many ways to meditate. The purpose of meditation is simply to develop a sense of awareness, focus intentionality (see below), insight and a selfless attitude. Meditation allows people to uncover, recover and discover parts of themselves of which they have not previously been aware. Meditation can be a bridge between the inner and outer worlds.

There are a great many types of meditation – lying down meditations, sitting meditations, standing meditations, walking meditations. There are breath meditations, sound meditations, visual meditations. Some are achieved through the intellect (one most popular with our society), through the emotions (used most frequently by mystics), through the body (as in yoga or dancing), or through action (as in concentrating on a particular skill such as rug making or perhaps gardening) to make it perfect. There are structured meditations and unstructured meditations. TT is sometimes called a healing meditation. This was because when subjects were attached to EEG machines, the EEG of a 'meditator' and TT practitioner were similar (Peper & Ancoli 1976).

EXERCISE 7 INTRODUCTION TO MEDITATION

You might wish to try different kinds of meditation until you find one that suits you. Whatever your choice, you should feel better for it. Try it for 10 or 15 minutes a day to begin with. Many books, tapes and so on are available as guides to different techniques. A good example is that by Le Shan (1974).

Below are a couple of short, simple meditations you may try out. Before you begin make sure that you won't be disturbed – use a quiet place where you will not be interrupted by others, telephone calls etc.

- Sit quietly, feet on the floor, eyes closed.

- Become aware of your breathing.

- Breathe slowly in and out, counting each time you breathe in, up to four.

- Just be aware of the counting – as fully aware as possible.

- When your mind wanders or sensations or emotions arise, bring your attention back to the counting.

The goal is to have your whole being involved in the counting.

Try this meditation again with eyes open and add 'and' to each count as you breathe out, i.e. one-and, two-and, three-and, four-and, one-and. These are only variations.

Start out with 10 or 15 minutes a day. Set your alarm clock so you don't have to look at your watch. You might want to put the clock under a pillow.

HEALTH PROBLEMS RESPONSIVE TO TT

Research has shown, as discussed in Chapters 1 and 2, that TT can accelerate wound healing, induce physiological relaxation, decrease anxiety, decrease pain, decrease diastolic blood pressure, and reduce stress in hospitalised patients. Clinical practice has shown TT to be particularly useful for upper respiratory tract infections, allergies, headaches and musculoskeletal conditions as well as being effective on an emotional and sometimes spiritual level. It is extremely useful for promoting comfort for those who are dying. It may not 'cure' the illness but it will hasten healing and help the body to heal itself faster.

There are some patients/clients with whom it is wise to proceed with caution and limit your time. These are the frail, pregnant women, babies, people with head injuries, emaciated people (AIDS or cancer patients) and patients with mental health problems. This is not to say that TT is ruled out, rather to emphasise the need to be especially aware and sensitive during the interaction.

TT WITH FAMILY MEMBERS

TT requires the practitioner to be objective, and in a place of non-attached compassion. Being objective is sometimes hard with family members or good friends, and it is often difficult to remain centred enough to be effective with them. However, it seems a shame that something so helpful cannot be shared with those we love most. So give it a try! It will be excellent practice for staying objective and centred. Remember though, if they are very ill and you are upset, it may only add to their discomfort.

INTENTIONALITY

Other than centring, the most important quality necessary for TT to be a positive act for both of the people involved is intentionality. The practitioner must have a strong motivation to help or heal. Intentionality implies not only the will to help or heal but also a specific goal in mind, i.e. the practitioner understands how to facilitate the interaction. At the same time, it is necessary to give up the expectation of a particular change happening. (See the note below on compassionate non-attachment for further explanation.)

The identification of intent is at least as important as the focus of the intent – perhaps more so. That is, in order to practice TT successfully, practitioners must acknowledge their motives. While intention to help patients achieve their maximum state of wellbeing is ideal, practitioners are also said to profit from trying to understand their desire to help.

In Rogers' (1970) model intent is addressed under her concept of acasuality. In this model everything and everyone – all energy fields – are interconnected. What each of us does directly affects everyone else. Therefore, the results of what one does is affected by other conditions, e.g. the other people involved, the energies around, the environment in which it's done. So it is possible to have intent to do something and yet for the effects to be quite different. The intent in playing billiards may be to shoot the ball, have it hit two others and have all three go into the pocket – but the results can be anywhere between that and none of them going in. The intent in turning a patient every hour may be to prevent pressure sores, but they may still occur. In TT the intention may be to see the person as whole through the repatterning of energy fields, but may produce something quite different. It is important to keep intention clear and then let go of the outcome. Inherent in this concept is the idea that by our thoughts we influence the outcome of any interaction. One of the basic tenets of quantum physics is that we are not discovering reality, but participating in the creation of it.

COMPASSIONATE NON-ATTACHMENT

Practitioners sometimes have difficulty in understanding the difference between intentionality and non-attachment. They find it confusing to have an intention and yet let go of the outcome. The following case study addresses this confusion:

This experience was profound for me and I always think of it when I get personally attached to a certain outcome. No one can assume that they know what the outcome of TT will be. We can only visualise the patient whole and administer TT to them from a compassionate, non-attached place.

In the hospital in which I was working I was a primary nurse for a man I shall call Bill. Bill had lymphoma and had been a patient of mine for over a year – coming in for chemotherapy every few months. He had been a professor of physics at the University and was extremely interested in my experiences while learning TT. He always initiated a conversation around this subject whenever he was admitted, yet when I offered to do TT with him, he would always decline. However, his obvious interest in what I was doing and indeed his confirmation of the things I was experiencing (from his physics background) was of great help to me. This was in the middle/late 1970s before I was aware of Rogers' framework.

One day I went to work and found Bill had been admitted and was in an acute and critical condition. As the shift wore on he became short of breath, his vital signs were not good and he developed a great deal of pain. While a decision was being made about his transfer to ICU, an X-ray was taken. He then said to me, 'you use TT to ease the dying, don't you?', to which I answered with a nod of my head. He then asked me to help him to die. On the assessment I felt a blockage over the heart and intense, uneven energy in his forehead. I began to do TT with Bill, holding the intent in my mind of easing the transition for him. 30 minutes later I felt his field to be balanced, but he was actually feeling much better and his vital signs were improving. I stayed with him for a while longer and then said goodbye as I had gone over my shift and was due back to work in just a few hours. However, neither one of us expected to see each other again. Coming back into work 8 hours later, I was met at the desk by a house officer who had in his hand two sets of X-rays – one which was taken prior to TT, one taken after I had left. The first X-ray showed a pulmonary embolus to his left lung, the second X-ray was clear of the embolus. No medical intervention had taken place. Bill had not been sent to ICU and he was looking much better. Ten days later, he died, but not before he had called to his bedside a sister from whom he'd been estranged for many years and with whom he was then reunited.

TT AND THE BENEFIT TO THE PRACTITIONER

The practitioners of TT may find themselves receiving immediate benefits from the relaxation and sense of wellbeing that comes from centring. In this state of consciousness, practitioners may experience unconditional love and compassion. McClelland's (1985) research has shown that when people feel unconditional love or compassion it may have an added benefit by increasing the effectiveness of the immune system of the practitioner. If this is so, then this offers an explanation why practitioners of TT often report feeling better themselves after practising TT and seem to have less illness (Quinn 1993).

'PROTECTION'

One of the questions most-asked by students of TT is how to 'protect' themselves. When questioned what the protection is for they usually mention 'negative' or 'bad' energy from the patient, outside forces that might interfere with the process or take one over, and even 'evil spirits', or 'the devil'.

Because the act of TT requires the practitioner to become sensitive to the energies of the other person (who is usually in some distress or imbalance), the practitioner may feel the mood of, or the pain of the patient if they are not centred. If, however, practitioners are able to maintain their own centre of consciousness, they will recognise that this mood or pain is not their own. Though they may need to spend a moment in recentring, they will quickly get back to the process itself – that of repatterning or rebalancing the energy field of the patient into a synchronous flow. If they lose their intentionality and centre and become identified with the patient and their problem, they will probably become part of the problem.

Practitioners find that moods are infectious. A joyous person quickly makes others happier. We 'pick up' sadness when we enter a room – or anger. Practitioners find that being in the presence of someone who is depressed can sap them of their energy very quickly. As they begin to be aware of energy fields of others, and how the interaction of fields affects them, they recognise that in the act of TT projections, transferences, countertransferences, depressions, anxieties (and likewise joy and feelings of well being) are greatly intensified. They realise the importance of clear judgement about their own inner resources and ability to be of therapeutic value to others. Therefore centring is again of the utmost importance. When practitioners are calm and strong inside themselves and can feel no personal investment in the outcome of the interaction, they are best able to set limits on how involved they want to become. Practitioners who are clear within themselves, centred and at the same time compassionate, with a deep intention to help or heal the whole patient, will not become adversely affected by the interaction (e.g. feel sapped of energy, pain in their own body or emotional distress). In other words, they will be 'protected'.

ONE STUDENT'S SOLUTION TO TAKING ON THE PATIENT'S PROBLEMS

I had problems with tiredness and absorbing the client's excess energy during the first session. I could not believe her pain that I carried away with me. The weight and vice-like pain made me feel very muggy.

So I had to review my centring. Initially, I felt that I shouldn't have to keep centring myself during the treatment session but after my experience in the first session, in the second session, I stopped frequently to recentre. That was helpful but didn't cure it, so I tried another exercise which proved more effective.

However, I was still absorbing the patient's congestion to a lesser degree. I tried to shake the excess energy off my hands frequently and to wash my hands and have a drink of water after the session. But I felt the problem was in my method of using my hands in combination with my breathing. I was very sure that I was absorbing my patient's energy when I breathed in as I was working with them. So I tried to take both hands away whenever I breathed in, but this made the treatment disjointed and I often forgot to do it, in any case. I remembered my TT teacher saying how we give energy with our right and receive with our left. THE ANSWER! I changed my technique to closing my left hand whenever I breathed in during the treatment session. It worked! I am now giving TT without getting tired or absorbing congestion.

Sometimes, before the TT interaction, practitioners will include in the preparation an image of the interaction being surrounded by white light or some other similar image. If this makes the practitioner more secure then these exercises are encouraged. Those who still feel uncomfortable and in the need of 'protection' after developing all of the above would probably do better practising another therapy.

Although people have been using touch for many thousands of years as a means of helping and healing, we are just beginning to understand on a scientific basis what happens when two energy fields come together. Although our energy fields interact with and affect others all the time, everyone who is drawn to give TT would agree that they wish to do no harm. Even though harm can be done – usually on an emotional or spiritual level – by self-professed so-called 'healers', the only uncomfortable side-effects to have resulted from 'Krieger's Krazies' (students who have been taught by Krieger), or from students of the Didsbury Trust in the UK, have been irritability, restlessness or sometimes an increase in anxiety. This sometimes manifests as actual pain felt by the patient in the area on which one is working, or as unexplained hostility coming from the patient. Krieger (1993) has noticed the same thing and calls it an 'energy overload'. It has happened in beginner's classes when many people are exchanging experiences. An inexperienced healer who is trying too hard, or someone who is trying to impress may also have this problem. It can also be observed when the practitioner begins the treatment from an overzealous state, which might arise from impulse, not intuition. In this case the practitioner is not acknowledging the intimate and emotional depth of the interaction nor being sensitive to the patients and their needs. Discomfort has also been caused to others by someone who is not centred. As in all energy interactions, the more sensitive the patient, the more care one must take.

> A good rule of thumb, I believe, is if you're unsure, don't. Do less rather than more. In TT you want to rebalance, repattern and modulate the patient's energy field into a synchronous flow, taking your cues from the assessment you make. No more, no less.
> TT Practitioner

Confidence is important in the act of TT. Confidence comes from knowing just how you are going to proceed and that you have the knowledge to deal with whatever may arise. This comes with practice and a continuously expanding knowledge-base. TT looks very simple and indeed it is. But what is really happening is hidden from view. TT is a paradox: though simple it is one of the most complex of all energy interactions, and practitioners need to spend a lifetime working on themselves and gaining understanding of the concepts behind it.

Factors affecting success

TT is obviously not the therapy that all nurses will choose to practice, nor is it the therapy of choice for all patients. Practitioners will get an intuitive sense of those who might respond, e.g. get a relaxation response and pain relief from TT. Occasionally this intuitive sense is wrong, but this is usually covered by asking permission of the patient. Asking for permission should be kept as simple as for instance, 'You seem a little tense', or 'it's not quite time for your pain medication. I've learned a technique that might help you. Would you like to try it?' Students can create problems if they start trying to explain the theory behind TT or engage in a discussion of the new physics with their patient. All the patient wants is relief from discomfort, and discussion or explanation of the theory may or may not come later on in the relationship. (Informed consent and legal aspects of TT will be discussed in Ch. 5).

Having suffered with mouth and throat pain for the past 8 years, with no relief from various medications, Dee was extremely distressed about her problem, for which no medical reason could be found.

Dee related how she believed that her problems started after having dental treatment. She experienced an unpleasant taste in her mouth at the time of the treatment but other symptoms then emerged, including numbness of her lips, pain likened to bee stings, loss of taste sensation and affected hearing.

These symptoms had a profound effect on her wellbeing: they caused sleepless nights, a reduced appetite and eventually the feeling that others thought she was neurotic.

Prior to starting the treatment, I outlined the basic theory behind TT. I explained that we are energy fields and that an interaction would occur between our fields, with the intention of repatterning and rebalancing her energy field. This at present was not flowing freely due to her pain and distress. Dee was intrigued by this theory and expressed her belief that there was 'more to life than the eye can see'. I further explained that the therapy itself involves me passing my hands over her body from head to toe in order to assess, rebalance and repattern her energy field.

Whilst Dee was making herself comfortable, I centred myself before commencing TT. This stage of the process is very important, because if one is not centred there is a danger of one's own energy becoming drained. It is then that one becomes focused on the intention to help or heal by finding a place of inner calmness. To help achieve this focused, alert and open state of mind, I visualise an oak tree shimmering in the breeze, with its roots going deep down into the earth.

The initial assessment of Dee's energy field was then carried out by moving my hands, a few inches above Dee's body, from head to toe in a continuous and flowing manner. This revealed congestion of the energy field as her head, neck and chest areas were 'cold'. Energy was therefore directed to these areas in order to loosen the congestion, to enable the energy field to flow smoothly. I visualised a warm soothing light travelling through me and into Dee and the congestion dissolving into this and then being taken out of the body through the feet. When I felt warmth in these areas, I then reassessed the whole energy field to pick up any new cues before rebalancing was carried out. This procedure was also carried out over the back area too, with no specific cues detected. When I had completed the treatment, Dee was obviously more relaxed and stated that she had felt warmth travelling from her head to her toes.

On Dee's second visit to the clinic a week later, she told me how she had been pain-free for a few days and had been able to sleep for a couple of nights without waking with pain. She was obviously delighted with this outcome of the TT treatment.

The second assessment of Dee's energy field revealed coolness over the areas where there had been a cold sensation the previous week but a deficit was felt over the throat. Energy was given into the cool areas as previously described, until a warmth was detected. The deficit was detected by a pulling sensation. Energy was given to the area until the pulling sensation ceased. Dee could feel an unusual sensation in this area but it did not distress her. Reassessment, balancing and repatterning continued until the body's energy field felt complete.

Dee visited the clinic on three more occasions. No more deficits were detected and the coolness over the head, neck and chest seemed to shrink in size as each TT treatment was given. Each treatment consisted of the process of assessing, directing energy and rebalancing and repatterning the energy field, each treatment ending with intuitive knowledge that the treatment and energy field were complete for that week.

As the weeks went by, Dee continued to experience more and more relief from pain. She was able to sleep at nights and able to enjoy her food once again, now that she could taste it. From being an anxious-looking person, who thought she would have to 'put up with the pain', for the rest of her life, Dee was now a happy, smiling woman.

After the practitioner has offered TT to a patient and they have agreed, there are other factors to consider which determine what kind of results will be achieved.

These include:

Condition of the patient

What is their diagnosis, prognosis? How long have they been in a state of imbalance or disease? What is their physical condition, emotional state, spiritual awareness?

Patient's expectation

Do they want to be different than they are? Do possible secondary gains through illness or pain outweigh the benefits of being different? For example, does the patient have some unwillingness to get well, perhaps because they do not wish to lose the regular visits by their nurse?

The practitioner's degree of expertise

Although almost anyone can learn the technique of TT in just a few minutes/hours and, with further teaching, confidence and practice the interaction will deepen and expand. It may take many years to become an expert practitioner. Some researchers have discounted studies where the practitioner had less than 3 years experience. Practitioners should not be discouraged by this – the best is yet to come.

Joe was a young engineer who had been admitted to hospital because of a fungus infection. It had started between his toes, then spread to the top and bottom of his foot. His toes were swollen and there was a strong smell emitting from the raw, red areas. He also had a severe rash, due to the side-effects of the antibiotics, which spread from his groin to his stomach and a red itchy rash on his knees. He was barely able to walk and had calf-strain due to poor mobility.

After a simple explanation of TT, I stood at Joe's right hand side and centred on my breathing. I was aware of feeling very peaceful and sensitively aware of Joe's presence. I thought of him as being well and happy. I experienced a deepening sense of quiet, and lost awareness of all external noises and events. I felt purposeful and held the intent within me to help and heal. I felt deeply centred. I was now ready to assess Joe's energy field. Using long, sweeping, flowing movements from head to feet, my hands receptive to the cues I was receiving, I made a mental note of the areas where his field felt imbalanced. The left side of Joe's head and face felt hot, the neck felt difficult to pass my hands over. I felt pressure when I was in the area of the liver and the stomach. At the lower abdomen my hands felt a tingling sensation which I also felt on his right foot and the last three toes of his left foot. When I assessed his back, I again had difficulty in passing my hands over the area of the liver, and a pulling sensation was felt in the middle of the back. The tingling over the right and left foot were the same. I cleared Joe's field using long sweeping movements of my hands. Continuing to feel my field interconnected with both Joe's field and the environmental field, I then projected, directed and modulated these energies until Joe sighed and looked relaxed. His breathing was also slower. I visualised the heat being dispersed from the side of Joe's face and head. I used short movements of my hands to change the energy imbalance, which I perceived as a dragging sensation, into one of harmony. My hands became very sensitive when working over the areas where I had sensed tingling. It was like receiving many tiny electrical impulses all across the palms of my hands. I continued modulating the energy field until Joe's field felt in balance.

Joe rested for 15 minutes and then he made comments such as 'I'm sleepy and my foot feels great. It's quite extraordinary'. He later wrote on the diary sheet I had given him: 'During the treatment I felt warm sensations to various areas, which had been sore, itchy and watery. Near to the end of the treatment I could visualise, in my mind's eye, shimmering shapes which reminded me of net curtains being blown by a breeze in a darkened room. It felt as though objects were being placed on areas of my body during her treatment. After TT, I felt relaxed and contented.'

When I saw Joe for his second TT session, he reported that he was now experiencing no pain in his groin area and that he felt the rash 'had been responsive to TT treatment'. I followed the same phases as I had in the first. The left side of his head and face now felt balanced. His neck area was still slightly heavy which I modulated energy to disperse the congestion which I felt there. The groin area was considerably more even, but over the right foot my hands still felt a tingling sensation. I repatterned his whole energy field and, when I felt the field was more in harmony, I brought TT to an end.

In the third session, the only area where my hands felt an imbalance was in the lower abdomen and right foot. I used long sweeping movements over the lower abdomen, down the legs and onto the feet. This cleared the imbalance in the lower abdominal region. I then held my hands over his right foot for a few minutes visualising Joe as well and happy. I felt an increasing sense of peace and 'at-one-ness'. The feeling of unity with Joe deepened. It is hard to put into words the tranquil silence, where time seems to no longer exist. The quiescent moment of totality. As I passed my hands over the front of Joe once more, I felt that the TT session was completed. Joe began to snore. I sat by his side, still feeling an overwhelming calm.

The practitioner's willingness and ability to confront self and motivation

The act of TT can look deceptively simple to an onlooker, it may look just like someone touching another person. In one of the research studies (Quinn 1982) 'mimic' TT was done as well as non-contact TT. Mimic TT is defined as an intervention that imitates the movements of the practitioner, but during which time there is no attempt to centre, no intention to assist the patient, no attuning to the condition of the patient and no exchange of energy. Both were videoed and the unpracticed eye could not tell who was doing which. But the fact of the matter is that a great deal is happening inside the practitioner. Anyone can be shown how to make the hand movements associated with TT but, without the intention and motivation to help and heal, the results of TT are diminished or rendered ineffective. The practitioner's motivations in TT – whether to offer sincere help and healing to another or for some other benefit (e.g. acting out a need to have another dependent on them, or for the sole purpose of making money) (Dass & Gorman 1986, Snow & Willard 1989) – will affect the degree of success of an episode of TT. People learn and apply TT for many different reasons, and part of the process of becoming an effective practitioner is to confront and examine motivations. Looking into the motivation for wanting to help will be a lifelong process, but it is necessary to make a start on this inner journey in order to become the kind of practitioner that professional practice demands. This exploration must be gentle and without guilt. It need not be shared with anyone, but it sometimes helps to share insights with a counsellor/supervisor or a trusted friend or colleague. Writing insights down in a journal can also be helpful. As more and more nurses and others learn about TT, it will become easier to find colleagues to offer support and clinical supervision. Practitioners are far more likely to become and remain effective when they can seek guidance and help from others as well. Those who care for others need someone to care for them too.

REFERENCES

Borysenko J 1985 Healing motives: an interview with David McClelland. Advances 2(2): 29–41

Dass R, Gorman P 1986 How can I help? Ryder, London

Krieger D 1979 The therapeutic touch: how to use your hands to help or heal. Prentice-Hall, Englewood Cliffs, N J

Krieger D 1993 Accepting your power to heal. Bear & Co, Sante Fe

Krieger D, Ancoli S, Peper E 1979 Searching for evidence of physiological change. American Journal of Nursing 79(4): 660–663

LeShan L 1974 How to meditate. Bantam, New York

Lionberger H 1985 An interpretive study of nurses' practice of therapeutic touch. Unpublished PhD dissertation, San Francisco, University of California.

Macrae J 1987 Therapeutic touch: a practical guide. Penguin, London

Peper E, Ancoli S 1976 Two endpoints of an EEG continuum of meditation. In: Krieger D 1979 Therapeutic Touch. Prentice Hall, London

Quinn J 1982 An investigation of the effects of therapeutic touch done without physical contact on the state of anxiety of hospitalised cardiovascular patients. Dissertation Abstracts International. University Microfilm No. DA 82-26-7882

Quinn J 1993 Psychoimmunologic effects of therapeutic touch on practitioners and recently bereaved recipients: a pilot study. Advanced Nursing Science 15(4): 13–26

Rogers M E 1970 An Introduction to the Theoretical Basis of Nursing. Davies, Philadelphia

Sayre-Adams J 1993 Therapeutic touch – principles and practice. Complementary Therapies in Medicine 1: 96–99

Sayre-Adams J 1994 Therapeutic touch – a nursing function. Nursing Standard 8 (January 19): 25–28

Snow C, Willard D 1989 I'm dying to take care of you. Professional Counsellor Books, Redmond, W A

5

Developing the practice

Jean Sayre-Adams   Steve Wright

The need for preparation
Consent and confidentiality
Deepening our understanding

■ *We shall not cease from exploration*
And the end of all our exploring
Will be to arrive where we started
And know the place for the first time.

T S ELIOT *Little Gidding*

Nurses and other health care workers have been caught up in the explosion of interest in complementary therapies in recent years. Some see this as a great opportunity to expand the boundaries of health care, while others fear the uncontrolled application of (not to mention the many 'unknowns' which surround) these therapies. All, including TT, have many potential benefits, even though their mode of operation is often hard to define or substantiate in strictly rational or scientific terms.

The telephone message was from the senior staff nurse on one of the adult oncology wards which use and support complementary therapies. The staff nurse is a qualified masseur, one of three who regularly treat the patients, but she had been unable to help the patient this time, and the physiotherapist would not attend without a doctor's referral. The message asked if there was anything that I could do to help urgently?

When I arrived, I asked the nurse for an update of the situation. It had begun as a simple movement of the head and a 'twinge', that ended in a spasm of torticollis. Now approximately 2 hours since the first movement, the patient was in 'excruciating pain and feeling helpless' as he curled further over the arm of his wheelchair in an effort to relieve it. I knelt in front of him covered his clenched fists with my hands in greeting and looked into his painfilled eyes. He was ashen, the fine diaphoresis covering his face and hands making his skin look even more transparent. The pain had rendered him helpless and vulnerable, he was unable to move, read, talk or swallow, and this meant not being able to take oral analgesics.

His scalp felt damp and cool on the surface, but by holding his head gently, heat could be felt, mostly on the right (twisted) side. Standing behind him, I placed my left hand on his forehead and gently and slowly massaged my right hand into the angle of his neck. As I held his head, I became aware of a deep, fine vibration in his neck. I asked him to breathe as slowly as possible and to allow the side of his neck to 'give' as it relaxed. I synchronised my breathing with his, allowing the energy to flow as we breathed out.

Gradually, his neck relaxed and the tissues under my hand changed from tense and hot to softer and more even. With an audible sigh, his neck relaxed and he was able to look at me with a straight gaze. He smiled and said 'I would never have believed it possible if I hadn't experienced it for myself'.

> Once back on his bed he was made comfortable and I completed the clearing which I felt was needed. As I continued from head to foot, the energy became more uniform in temperature and flow. Each time I returned to his head he would open his eyes and smile, following my hands with his gaze until his eyes closed and stayed closed as he surrendered into a restful, painfree nap.

The need for preparation

When enthusiasm for a particular subject develops, one of the risks is the tendency to fall into 'quick fix' ways of learning about it. The proliferation of courses in complementary therapies extends from modules included in Masters Degree programmes and in-depth courses validated by reputable bodies (such as nursing's National Boards) to 'one-off' seminars and teaching sessions of dubious value, led by teachers with even more dubious qualifications. Unfortunately this plays into the hands of those who seek to vilify practitioners in complementary therapies as quacks and charlatans. Earlier chapters have suggested that there is a growing research and theoretical basis to support the practice of TT, and that it can be learned by anyone. It requires no special equipment and is not, at least in conventional terms, an invasive technique. However, the apparent simplicity of the practice of TT belies its fundamental complexity and intricacy. Thus, in Chapter 4 we have argued that no one should practice TT with patients unless they have satisfactorily completed a recognised programme of preparation as a practitioner.

Courses which teach TT are, however, only part of the commitment which nurses and other health care workers need to make. In Chapter 4, it was clearly identified that the success of TT depends not just on mastering the theory and practice, but also in the practitioner's continued development of personal and spiritual growth, awareness, assertiveness, expertise and so on. The conclusion was drawn that practitioners are best able to offer holistic help and healing to others, when they are themselves learning about and developing themselves as well as their practice. Furthermore they need to work in a culture which encourages and supports them in their skills.

The concept of clinical supervision, for example, is now strongly advocated for health care professionals (Butterworth & Faugier 1993). As the numbers of practitioners of TT expands, so too will the opportunity for them to seek clinical supervision by experts in their own field. The Didsbury Trust (Sayre-Adams 1993) is currently offering support in this respect, and strongly advocates that all TT practitioners should have access to a clinical supervisor with expertise in TT.

Currently the Didsbury Trust is the only body in the UK which offers a variety of courses, from introductory 2-day seminars to longer programmes validated by the Welsh and English National Boards for Nursing, and the University of Manchester. The course designed for practitioners lasts a year and includes formal teaching sessions, the use of reflective practice, clinical supervision, case study and dissertation work.

Students undertaking a TT course frequently find it to be a transformative experience. The sense of learning and excitement in their evaluations is almost palpable.

REPORTS FROM STUDENTS OF TT CLASSES

'The energy literally sizzled in the space between my hands and his palms. It is all so special and yet so ordinary'.

'At a time when before I have felt rather helpless and would possibly have avoided the patient, I now feel that there is something to offer, not only to the patient, but it may also offer comfort to the relatives. By showing them the basic TT techniques, it gives them a means to communicate and it may be that this helps in the commencement of the grieving process'.

'There is the personal pleasure of having learnt a new skill, plus the stimulation and excitement that comes with satisfaction as new knowledge is acquired. The information gained from being able to offer a new concept can only bring benefit to existing nursing practices'.

'Like a bird on the wing has an expanded vision, so Therapeutic Touch has expanded my horizon'.

'My whole horizon has been opened up by the experience of TT. This in turn has been reflected in my work, both in my written work and my practical work and also of course in everything I do. It's almost like being a new person, certainly a wiser one'.

'I am more grounded than I used to be.'

'The deeper I enter this awakening within, the more difficult it becomes for me to express it in words. I feel as if I have found myself.'

'Without emotional attachment but with a loving care I can see my weaknesses and strengths, and this is helping me to grow, because I can accept me as I am. Knowing that I am not perfect, and that this is part of me, helps me to reach others because I too have lumps and bumps and cracks and wrinkles that are part of me, and makes the wholeness real. Before, I had great difficulty in acknowledging these parts. They were too painful, too hateful, too mortifying. Now they are the light and dark that an artist has used to get the effect of depth and realness.'

'One of the most rewarding aspects of my work is to show parents ways of coping with the stresses of family life. There is often that feeling of helplessness when caring for a crying, fretful or sick child, which soon can lead to loss of confidence in themselves as parents. By teaching the simple technique of TT to parents, they then feel they have something special to offer, which in time increases their self esteem and so strengthens the bond between them and their child'.

The personal growth and pleasure which students may get from a TT course, however worthy in themselves, are not sufficient to substantiate it in practice. Practitioners need to be safe, not only to maintain the credibility of a particular therapy but, most importantly, to maintain public safety and confidence.

Many professional organisations require practitioners to demonstrate competence in any technique they use in their practice. In the case of nursing, for example, a Code of Professional Conduct (UKCC 1992a) is very explicit about the goals of nurses to 'act, at all times, in such a manner as to safeguard and promote the interests of individual patients and clients'. Each nurse is 'personally accountable' for his or her practice, and may be subject to disciplinary action, even removal from the register of nurses (meaning that he or she will no longer be able to practice as such).

Thus nurses are required to demonstrate that their professional knowledge and practice is up to date. When they add new dimensions to their practice (such as learning about working as a practitioner of TT) they must do so only if they can demonstrate adequate preparation which makes them competent to offer such a service. Indeed, organisations such as the Royal College of Nursing, which includes indemnity insurance in its membership fees, make the same requirement of their members if they are to maintain their insurance cover. Fortunately, litigation by dissatisfied patients against nurses is still relatively rare, and it is important to keep the risks in proportion. However, when nurses ask whether they are covered to give TT, one of the factors which govern the response is whether or not the person offering it can demonstrate competence through properly validated training. One of the common concerns registered by critics of complementary therapies is the lack of controls over who can practise them. Where practitioners are governed by professional Codes of Conduct, this at least appears to offer the public protection. The UKCC (1992b) recognised that nurses expand their practice in many ways beyond initial graduation and registration, and reinforced the personal accountability of nurses when they do so. It recognised that expecting nurses to hold a 'certificate' for every additional practice they undertake to be overly simplistic. Instead, it emphasised adherence to a number of 'principles of practice', including the need to maintain and develop knowledge and competence underpinning any aspect of nursing practice in order to promote patient safety.

Consent and confidentiality

The maintenance of patient confidentiality and consent to treatment with TT is governed by the same principles applicable to any other nursing or medical intervention.

A 23-year-old woman with whom I work was in the beginnings of a migraine. I offered to do some TT with her. As she agreed she told me she was uptight about her situation and wanted help to relax and perhaps some feedback as to how she could take responsibility to make changes in her life that would help her avoid getting in the state in which her migraines occurred.

I began treatment with the patient lying supine on the bed – she did not want to turn over, but preferred to remain on her back throughout the treatment. I carefully supported her arms and the area underneath her knees with pillows to allow her position to facilitate maximum relaxation.

I began by explaining how we affect one another's energy fields and that I may work with my hands in contact with her body sometimes and at other times hold my hands a little way from her body. I told her that her discomfort was likely to start to subside and continue to do so after treatment if it was not completely gone at the end of the session. I also asked her to raise her right hand, if she wanted me to stop the treatment for whatever reason, and then tell me what her needs were in her own time. I told her that she could help with her treatment by picturing a 'letting go' of her discomfort and seeing herself in full health. I prepared her to expect to lie quiet for 20–30 minutes after the session.

During the centring I stood at the head end of the patient with the top of her head facing me and stayed this way for about 5 minutes. She gradually relaxed, her facial expression became softer and her breathing deepened and slowed.

If I start 'thinking' and 'planning' what I am to do, beyond holding the general framework in the background of my mind, I become self-conscious and end up doing something *to* the patient and get attached to the results. The 'centring' process is about tuning in to the 'intuition' and letting whatever comes flow. When my 'self' gets out of the way there is a deep inner 'knowing' that I touch, a process on a deeper level. Words are very clumsy tools for expressing this process.

When she seemed really relaxed, I did the assessment on her.

I felt areas of congestion, heaviness and blockages. There were some cooler areas as well as emotional tension and conflict.

I had a sense of the left side absorbing energy, the right side felt stagnant. I worked in an anticlockwise circular movement until the energy was flowing more and more freely. Next I worked with my hands about 4–6 inches away from the body and 'smoothed' the whole energy field from head to toe. The main part of the treatment involved holding my right hand fairly close to the left side of the throat area and the left hand fairly close to the face and crown. I had a picture of blue-white light flowing from my left hand and being directed by my right hand. After about 5 minutes I 'knew' it was time to stop and I completed the session with a quick check of the general energy field and balance of the feet. Some tears appeared while I held my hands near the neck area – there was a sense of something being released. The patient was almost asleep at the end of the session and I woke her up after half and hour's rest. She felt ready to get up and go.

Consent to any treatment may be 'implied', e.g. the patient in the hospital bed is presumed to have submitted him/herself to a reasonable range of care. Thus, every time a nurse pursues a particular activity with a patient, it is not necessary for a consent form to be signed. However, it is still open to debate as to how far TT can yet be considered part of mainstream nursing. In some settings, it may be necessary for patients, or their guardians, to be asked to sign consent forms when TT is offered, however this tends to be unusual. Experience suggests that most nurses and other health care workers are offering TT as part of their normal range of helping and healing activities. A more critical factor is the need for consent to be informed. As the above example illustrates, explaining what TT is all about to the patient beforehand is important, giving the patient an opportunity to ask questions and give or deny permission. Such explanations need to be in terms which patients understand and give realistic expectations of the effects.

An increasing emphasis is being placed on nurses and other healthcare workers working in 'partnership' with patients and clients. This trend away from the authoritarian ('I know best') professional and the passive ('I'm in your hands') patient is reinforced by recent publications such as the Patient's Charter. Of course, patients' responses to healthcare workers can be extremely variable. Some patients may wish the professionals to 'take control' while others will demand information and control over all elements of their care. It is important to be able to respond to all the different approaches along this continuum.

Empowering patients and clients by giving information is a significant feature. They are better able to be involved in the decision-making process and give informed consent if they know what is on offer, and they have confidence in the practitioner. The Complementary Therapies Forum at the Royal College of Nursing has produced guidance for users of complementary therapies to help them get the information they need. It suggests (RCN 1993) a number of questions which the user should ask:

- What are their qualifications and how long was their training?
- Are they a member of a recognised, registered body, with codes of practice?
- Can they give you the address and telephone number of this to check?
- Is the therapy available on the NHS?
- Can the GP delegate your care to the therapist?
- Is this the most appropriate complementary therapy for your problem?
- Do they send a letter to your GP advising him/her of any treatment received?
- Can you claim for the therapy through your private health insurance scheme, if you have one?
- Are your records confidential?
- What is the cost of the treatment?
- How many treatments should you expect to need (and therefore approximate cost)?
- What insurance cover does the therapist have?

Asking such questions will help to ensure that participation in TT is based on informed consent, and also reduce the risk of exposure to bad practice.

A gentleman came to the clinic suffering with a pain in his right foot which had been troubling him for some 18 months and also with a pain in his right shoulder and neck, which had been present for a few months. The injury to Mr W's foot had occured while on holiday. He had dived into a swimming pool and on doing so felt 'something go' in his heel and thereafter had suffered pain and stiffness in this area. This discomfort varied in intensity depending on how much walking he did and the type of shoe worn. The pain in his neck and shoulder had no explicable cause. Mr W is 57 years old and works as a clerk in a local public transport services department.

On Mr W's first visit, he looked very apprehensive as he did not know what TT entailed. I reassured him by explaining that all that was required of himself was to remove his shoes, lie on the couch and make himself comfortable. I further explained that I would simply pass my hands over his body in order to assess and rebalance his energy field, which I explained in simple terms. Mr W seemed satisfied with this explanation, with no more questions being asked.

After Mr W had made himself comfortable on the couch, I proceeded to centre myself by visualising an oak tree with its leaves shimmering in the gentle breeze and inwardly calming myself by taking a few deep breaths, focusing on the intention to help and heal.

Following this I carried out an initial assessment of Mr W's energy field by placing my hands parallel to each other, a few inches above his body surface and starting from the head, moved my hands down towards his feet, in a continuous flowing motion. Mr W had a distinct coolness over his right foot, which indicated a blocked area of tight congestion. When asked if this was the foot troubling him, he confirmed that this was so.

As this blockage was present, I placed my hands under the arches of Mr W's feet and directed energy into this area to penetrate and loosen the congestion. This I did for a couple of minutes. This was enough for the blockage to move and energy to flow evenly in both feet.

I then reassessed the energy field but did not pick up any new cues, although the foot still felt cool. Therefore, I continued to energize the foot in order to promote the circulation of the energy flow. When the area began to feel warm, I discontinued this approach and carried out further reassessment and rebalancing of the whole energy field.

Following this, I then carried out an assessment of Mr W's back, from head to feet. Again I felt coolness around the neck and shoulder, and energized this area and thereafter reassessed and rebalanced the whole energy field.

The treatment lasted for approximately 20 minutes and when I had finished, Mr W asked 'is that it?' Mr W had not felt any sensations, for example warmth or tingling, and obviously wondered what it was all about. I explained that not everyone experienced sensations but that did not mean no beneficial effect would be experienced.

While I carried out the treatment, I remained centred, focusing my attention on helping Mr W to be well and balanced. When rebalancing the whole field, I visualised warm light spreading throughout the body.

Although Mr W did not seem convinced that anything had been done to help him, he did make a further appointment for the following week.

When he arrived the following week, he informed me that almost as soon as he had left the clinic the previous week, the pain in his right foot had much improved and for most of that intervening week, he had been pain-free, although the pain in his neck and shoulder remained.

On that second visit, I carried out the treatment as for the previous week, energizing the foot and shoulder areas in order to relieve congestion gathered there and drawing it out through the extremities and giving an overall rebalancing. On this occasion Mr W did experience a warm sensation in his foot.

The third treatment was carried out as for the previous weeks but on assessment, the foot area definitely did not feel cool as it had on the first, and Mr W was very pleased in the improvement, as he said some days previous to treatment, he was unable to walk downstairs due to the pain. As the shoulder and neck remained painful, I spent more time on this area, directing energy to loosen the congestion as coolness was still felt in this area. I visualised a warm, light stream of energy penetrating a knot and slowly undoing it.

On the fourth and final visit, Mr W said he felt a fraud coming, as he felt so much better, the pain in his foot, neck and shoulder having been relieved. Even though he felt better I carried out an assessment of his energy field and it felt more complete with no cool areas being detected. The treatment was completed with an overall rebalancing of the energy field.

Unknown to me, Mr W suffers with arthritis in his left hip, which causes him pain and was now more tolerable having received TT.

Deepening our understanding

There are great shifts occurring in all of our ways of thinking and doing things as the end of a millenium draws near. Sometimes the changes seem too fast and too radical for the individual to feel comfortable with. However, many believe that we can replace the systems that no longer serve us and participate in creating environments that will support healing. In the preceding chapters, we have looked at the history of healing, of laying on of hands and of Therapeutic Touch (TT). We have seen the attitude of society to this healing art change from faith to fear, but as the contemporary shift occurs it is slowly gaining not only acceptance but also popularity.

Nurses are flocking to the healing arts as they long to put the heart back in nursing. They are interested in TT because they see it brings relaxation, healing and pain relief to their patients. There is a beginning awareness of the conceptual framework that is often used for TT and so it satisfies their need to know why. Since research was first begun on TT in 1975, we have seen large amounts of good quality research coming particularly from North America by nurses in their masters' and doctors' degrees.

The need for more research is evident, and is reinforced by Quinn (1989):

> *The first and perhaps most obvious (need) is the application of TT to a host of real world clinical problems –clinical trials or outcome studies. The second path involves the development of an explanatory model/theory of TT that is validated and refined through the research process'.*

A deepening of understanding around Rogers' field theory as well as continued and replicated studies on TT is needed. However, the evidence shows that interest in and commitment to TT has grown rapidly among practitioners for essentially one reason – it seems to work (Sayre-Adams 1993). TT offers an important dimension in healing and caring for many healthcare professionals. Research so far suggests that it can bring significant benefits to patients, at minimal cost. While helping some patients, no harmful effects have yet been demonstrated when done by practitioners who have been trained by Krieger or her students and follow the guidelines set up by the governing bodies of TT.

The Didsbury Trust in the UK is developing TT based on accredited and well-established programmes. The Didsbury Trust is teaching TT in Colleges of Nursing, post-basic education departments and Nursing Development Units throughout the UK. The first of the practitioner classes, accredited by the Welsh National Board, was completed in the autumn of 1993 and others are in progress. It is out of these practitioners' classes that the future teachers and researchers for the UK and the rest of Europe will come. While the use of TT continues to expand in the UK there is a parallel need for qualified teachers and a considerable body of research evidence to support its use and deepen its theoretical base.

The challenges that are being presented in the UK and Europe only mirror the challenges in the U.S. Professional nursing bodies such as the American Holistic Nurses Association and the Nurse Healers Professional Association are also talking about regulation and educational standards for those who practice TT.

There are two ways of looking at or conceptualising the place of nurses and other healthcare workers in the healthcare environment. One of them is seeing the carer and the patient as innocent victims of a decaying, worn out, ineffective system. The other is to see the nurse and the patient as conscious participants in the changing of the system. Barrett (1990) says:

Nursing care in this [Roger's] system is concerned with patterning the environmental field. The nurse, together with the client, patterns the environment to promote healing and comfort.

If nurses and other carers chose to look at our healthcare system from this second view, the kinds of questions that arise are 'what can we do to create a healing environment for the patient? How could the environment be changed?' When we turn our attention to these questions, the issues of colour, light, sound, activity, temperature take on importance. 'Regardless of the practice modality being used, the nurse's objective is to pattern the client's environment to promote health and well being' (Barrett 1990). Many nurses and others within the health care system have started to alter the environment by the use of lighting, imagery, relaxation, music therapy and art therapy. They promote the health and wellbeing of their patients by the use of massage, aromatherapy and TT and other complementary therapies.

However, considering the conceptual framework that incorporates theoretical perspectives suggested by Rogers (1970, 1986, 1990), expanded by Newman (1979, 1982, 1986, 1990) and tested by Quinn (1982, 1984, 1989b) and others, what of the idea that the nurse may become the environment of the patient? In this perspective, the nurse turns toward his or her understanding of the 'nurse' as an energetic, vibrational field, integral with the patient's environment. When the focus is on the nurse as the environment the questions that arise are 'How can I be a more healing environment? How can I use my consciousness, my being, my voice, my touch, my face, for healing?' (Quinn 1992). Quinn raises a profound question and voices the thoughts of many nurses when she asks 'What are the limits of influence if the means of influence is an energy field?' (Quinn 1992).

It is necessary to review the conceptual framework that Rogers introduced into nursing in 1970. This conceives the fundamental unit of the living system as an energy field, coextensive with the environmental energy field. Although at the time, this was a revolutionary idea in nursing, and an unacceptable conjecture of little or no interest to Western medicine, a quarter of a century later, one cannot approach the cutting edge of practically any modern scientific discipline, nor the tradition of any major spiritual culture, and not see the same idea proposed. No longer merely conjecture, the interconnectedness of all life seems clear. Scholars, artists, and futurists alike are writing about this phenomenon in fields as diverse and different from each other as quantum physics, biology, Western medicine, Ayurveda, psychology, psychoneuroimmunology, philosophy, theology, spirituality and nursing (Quinn 1992). The emerging view of our world includes the concepts that the human being is a nonmaterial, multidimensional field integral with the environment/universal field; that consciousness is nonlocal, unbounded by physical structure and function; and that separateness of the individual from all other individuals is an illusion. Western mystics like Hildegard of Bingen and contemporary artists like Alex Gray provide us with insight into these same phenomena through exquisite paintings, poetry, and prose depicting contact with the energetic nature and underlying unity of all of life (Quinn 1992).

HEALING ENVIRONMENTS

Taking these basic assumptions, we need to explore the potential of this model of reality for creating healing environments. A healing environment is one that facilitates the emergence of the *hælan* effect, the synergistic, organismic, multidimensional response of whole persons in the direction of healing and wholeness. Healing, the emergence of right relationship at, between, and among all the levels of human being, is always accomplished by the one healing. No-one and nothing can heal another human being. All healing is creative emergence, new birth, the manifestation of the powerful inner longing, at every level, to be whole (Quinn 1992). What is the role then for the nurse and other healthcare workers? The nurse can remove barriers to the healing process: as Nightingale said (1869), the role of the nurse is to put the patient in the best possible position for nature to act. The nurse can participate in creating environments that will support healing. The nurse can become a midwife to the process of healing, creating and being safe. The nurse can be the sacred space into which the healing might emerge. The nurse can, literally, BECOME the healing environment. Further, taking from these theorists and disciplines the basic assumption that we are inter-connected to all life, our consciousness is not separate and apart but integral with all consciousness. The nurse can become a conscious participant in the web of interconnectedness toward repatterning and healing for ourselves and for others through the intentional use of our own consciousness. The clinical practice of TT is an example of this participation: the first and most important step in TT is a shift in the consciousness of the practitioner (in the act of centring) through which, clinical experience and empirical study demonstrate, there can also be a shift in consciousness of the recipient (Quinn 1992). Cowling (1990) suggests that:

> ... the nurse could knowingly participate in human field patterning through his/her interconnectedness to the client, possible in a four-dimensional [pan-dimensional] universe of open fields. For instance, it has been suggested that a human-to-human field process operates in TT through the mode of ... intentionality on the part of the nurse.

Intentionality is an important concept in the practice of TT.

TUNING IN

What may actually be happening in the act of TT is that the practitioner, through the act of centring, is repatterning his/her own energy field in the direction of expanded consciousness, a consciousness experienced as unified, harmonious, peaceful, ordered, and understood to be a 'healing meditation'. It is proposed that being in the TT process with a nurse in such an expanded (meditative) state of consciousness may provide a template of sorts upon which the patient may repattern. Using the metaphor of sound, the pattern of vibration of the nurse's consciousness becomes a tuning fork, resonating at a healing frequency, and the patient has the opportunity within the mutual person-environment process to attune, to resonate, to that frequency (Quinn 1992).

This perspective involves a shift from previous (up to 1993) descriptions of the basis behind TT which utilise an 'energy exchange' model that includes a here and a there: a source of energy and a recipient of energy. In a model of interconnectedness, to which this text inclines, there is no here or there – there is just now and always. This is more consistent, for example, with Rogers' theories. The whole issue of timelessness that practitioners experience in the process of TT is explored by Newman (1982, 1990), Bentov (1977) and most recently by Quinn (1993).

The poet T S Eliot put it another way in *East Coker* (from *Four Quartets*):

In order to arrive at what you are not
 You must go through the way in which your are not.
And what you do not know is the only thing you know
And what you own is what you do not own
And where you are is where you are not.

There is a challenge for all nurses who aspire to practice holistically, i.e., who accept the basic premise of holism, of an interconnected universe, and of the fundamental inseparability of individuals one from the other, to become conscious participants in the creation of healing environments. In this view the systems are not 'out there'. Nurses ARE the environment, nurses ARE the systems they work in. By changing and expanding their consciousness, nurses can become the facilitators for change and expanded awareness for their patients, their colleagues, their community, and their planet. A series of studies on the effects of the number of people practicing transcendental meditation (TM) in various cities and the corresponding crime rates have been carried out. The data demonstrates decreases in crime rate when more than 1% of the community meditate regularly. This effect has been termed the 'Maharishi effect' (Quinn 1992), since it was predicted by the Maharishi Mahesh Yogi, the founder of the TM movement. What would be the effect in a single hospital if 1% of the nurses began to practice TT and other healing modalities using expanded consciousness?

THE HUNDREDTH MONKEY

Watson (1980) tells of a phenomenon that may capture a vision for the future of nursing.

The Japanese monkey, Macaca fuscata, has been observed in the wild for a period of over 30 years. In 1952, on the island of Koshima, scientists were providing monkeys with sweet potatoes dropped in the sand. The monkeys liked the taste of the raw sweet potatoes, but they found the dirt unpleasant. An 18-month-old female named Imo found she could solve the problem by washing the potatoes in a nearby stream. She taught this trick to her mother. Her playmates also learned this new way and taught their mothers too. This cultural innovation was gradually picked up by various monkeys before the eyes of the scientists. Between 1952 and 1958, all the young monkeys learned to wash the sandy sweet potatoes to make them more palatable. Only the adults who imitated their children learned this social improvement. Other adults kept eating the dirty sweet potatoes. Then something startling took place. In the autumn of 1958, a certain number of Koshima monkeys were washing sweet potatoes – the exact number is not known. Let us suppose that when the sun rose one morning there were 99 monkeys on Koshima island who had learned to wash their sweet potatoes. Let's further suppose that later that morning, the hundredth monkey learned to wash potatoes. Then it happened! By that evening almost everyone in the tribe was washing sweet potatoes before eating them. The added energy of this hundredth monkey somehow created a mass creative breakthrough! But notice, the most surprising thing observed by these scientists was that the habit of washing the sweet potatoes then jumped over the sea. Colonies of monkeys on other islands and the troops of monkeys on the mainland – in fact all Macaca fuscata monkeys everywhere began washing their sweet potatoes before eating them!

(Watson 1980)

As you turn from being a victim of the system into a conscious participant in the changes happening, could YOU be the hundredth monkey?

REFERENCES

Barrett E A M 1990 Health patterning with clients in a private practice environment In: Barrett E A M (Ed) Visions of Rogers' science-based nursing. National League for Nursing, New York

Biley F C 1993 Creating a healing environment. Nursing Standard 5 (October 8): 31–35

Butterworth A, Faugier J 1993 Clinical supervision and mentorship in nursing. Chapman & Hall, London

Cowling W R 1990 A template for unitary pattern-based nursing practice. In: Barrett E A M (Ed) Visions of Rogers' science-based nursing. New York,. National League for Nursing.

Newman M 1979 Theory and development in nursing. Davies, Philadelphia

Newman M 1982 Time as an index of expanding consciousness with age. Nursing Research 31(5): 290–293

Newman M 1986 Health as expanding consciousness. Mosby, St Louis

Newman M 1990 Newman's theory of health as praxis. Nursing Science Quarterly 3(1): 37–41

Nightingale F 1869 (1980 edn.) Notes on nursing – what is, what it is not. Churchill Livingstone, Edinburgh

Quinn J 1982 An investigation of the effects of Therapeutic Touch, done without physical contact, on state anxiety of hospitalised cardiovascular patients. Michigan University Microfilms, Ann Arbor

Quinn J 1984 Therapeutic Touch as energy exchange: testing the theory. Advances in Nursing Science 6(2): 42–49

Quinn J 1989a Future directions for Therapeutic Touch research. Journal of Holistic Nursing 7(1): 19–25

Quinn J 1989b Therapeutic Touch as energy exchange: replication and extension. Nursing Science Quarterly 2(2): 79–87

Quinn J 1992 Holding sacred space: the nurse as healing environment. Holistic Nursing Practice 6(4): 26–36

Quinn J 1993 Psychoimmunologic effects of Therapeutic Touch on practitioners and recently bereaved recipients: a pilot study. Advances in Nursing Science, 15(4): 13–26

RCN (1993) Complementary Therapies – A consumer checklist. RCN, London

Rogers M E 1970 An introduction to the Theoretical basis for nursing. Davies, Philadelphia

Rogers M E 1986 Science of unitary human beings. In: Malinski V M (Ed) Explorations on Martha Rogers' science of unitary human beings. Appleton-Century-Crofts, Norwalk, Connecticut

Rogers M E 1990 Nursing: science of unitary, irreducible, human beings: update (1990). In: Barrett E A M (Ed) Visions of Rogers' science-based nursing. National League for Nursing, New York

Sayre-Adams J 1992 Therapeutic Touch: research and reality. Nursing Standard, 6(50): 52–54

Sayre-Adams J 1993 Therapeutic Touch. Cited in Trevelyan J, Booth B 1993 Complementary Therapy. Nursing Times Sourcebook, London

UKCC 1992a The code of professional conduct. The United Kingdom Central Council for Nursing, Midwifery and Health Visiting, London

UKCC 1992b The scope of professional practice. The United Kingdom Central Council for Nursing, Midwifery and Health Visiting, London

Watson L 1980 Lifetide. London, Bantam

6
Conclusions

Jean Sayre-Adams   Steve Wright

■ *The day will come when,*
after harnessing the winds, the tides,
and gravity, we shall harness for God
the energies of love. And on that day,
for the second time in the history of
the world, man will have discovered fire.

TEILHARD DE CHARDIN

So far we have sought to identify the research and theoretical basis for TT, and the central issues in its practical application. A number of key points are worth summarising:

- TT is a harmless technique in which energy fields are repatterned into a balanced state, putting those involved in the best possible position for relaxation and maximum health to occur.

- The practice of TT can be learned by anyone who so desires.

- There is a need to carry out more research on TT, including rigorous identification of patient outcomes as the UK, and healthcare systems worldwide, focus more and more on 'cost-effectiveness' and 'value for money'.

- As more and more nurses and others learn about and practise TT, they will be able to contribute to expanding its theoretical underpinnings. At the same time there is an opportunity to draw upon relevant and related concepts from many other fields of thinking.

- Education for the practice of TT needs to be well validated using a wide range of enlightened techniques. The Didsbury Trust is already advanced in this field. Standards of education and practice should conform to the requirements of reputable bodies such as the National Nursing Boards and the British Complementary Medical Association. The British Association of Therapeutic Touch (BATT) was launched in the summer of 1994 whose aim is the registration and accreditation of TT practitioners.

- All practitioners must undertake recognised programmes and be registered with their own professional body, and have appropriate insurance cover. The protection of patients is paramount.

- Practitioners need continuous updating and access to clinical supervision. Where they work in organisations, such as the NHS, then a supportive culture and climate is needed which is conducive to health and the support of both practitioner and patient.

■ TT does not stop with the nurse – patient relationship. Its significance has ramifications beyond nursing and into the work of all carers, the healthcare system and society as a whole.

As interest in TT expands, it is important to examine it not just from the perspective of a new technique to be added on to the existing practices of nurses and other health care workers. Nor, as we have tried to suggest in these pages, is TT a simple linear relationship from healer to patient. There is a grander perspective to view – that of engaging in the intricate and dynamic torrent of universal energy. The potential, when thousands of nurses and others participate intentionally in this cosmic exchange, is to transform the health and wellbeing of millions of people. TT is not a panacea; it will not cure every patient nor will it cure all the health system's ills or help all patients love all their carers all of the time. It is, however, part of what may be seen as a worldwide movement for change. It changes those who participate in it, and they in turn may participate in changing the world. It is a piece of the jigsaw puzzle in the search for health that, when locked into place, helps us to appreciate the bigger picture, the awesome picture in which we find ourselves.

A USEFUL ADDRESS

British Association of Therapeutic Touch, 33 Grange Thorpe Drive, Burnage, Manchester UK M19 2LR

Index